Low Temperature Stress Physiology in Crops

Editor
Paul H. Li, Ph.D.
Laboratory of Plant Hardiness
Department of Horticultural Science
and Landscape Architecture
University of Minnesota
St. Paul, Minnesota

CRC Press
Taylor & Francis Group
Boca Raton London New York

CRC Press is an imprint of the
Taylor & Francis Group, an **informa** business

PREFACE

The importance of low temperature stress physiology research has become increasingly apparent in agriculture for productions of food, fiber, and ornamental plants. A severe winter freeze or untimely cold in a major production area can influence the world food reserve and affect economic and human health conditions. In many cases, the difference between a good crop and crop failure is determined by only a few degrees of cold tolerance. Such subtle differences should be highly amenable to a research solution. The advances being made in low temperature stress physiology offer new strategies for the study of how plants react to stress, and also provide new clues for the improvement of production systems that will attenuate losses and the development of new cultivars with enhanced tolerance to stress.

This volume consists of two parts: namely, responses of chilling-insensitive plants to low temperature stress, and responses of chilling-sensitive plants to low temperature stress. There are a total of 14 chapters including 6 chapters dealing with cold acclimation related topics, e.g., alterations of gene expression and of membrane proteins and lipids, 6 chapters dealing with freeze stress related subjects such as deep supercooling and membrane ATPase as a site of freezing injury, and finally 2 chilling related chapters of ethylene production and mefluidide protection of chilling sensitive crops. References cited are listed at the end of each chapter. It is organized and presented to be useful to graduate students, teachers, and researchers.

The editor thanks the authors for their cooperation and efforts in preparing this volume. Their contributions have added greatly to our knowledge of low temperature stress physiology in crops.

Paul H. Li

THE EDITOR

Paul H. Li, Ph.D., is currently Professor and Coodinator of the China Program at the University of Minnesota College of Agriculture in St. Paul, Minnesota. He received his B.S. degree in Horticulture from Taiwan National Chung-Sung University in 1956 and his Ph.D. in Plant Physiology/Horticulture from Oregon State University in 1963.

Professor Li devotes his full time in research, teaching, and advising in environmental stress physiology of horticultural crops. His research concentrates on aspects of temperature stress physiology of crop plants, such as, molecular biology of potato cold hardiness, mechanisms of freezing injury and resistance of tissue culture, adaptability of bean plants to high temperatures, and mefluidide chilling protection mechanisms in corn plants.

As a member of the faculty at the University of Minnesota, Professor Li teaches courses in Horticulture Science and in the Graduate Plant Physiology Program. He is also a full member of the Graduate School of the University of Minnesota advising M.Sc. and Ph.D. graduate students majoring in horticulture and plant physiology.

Professor Li is a member of Phi Sigma and Sigma Xi and is the recipient of numerous awards and grants. He is also a manuscript reviewer for numerous agricultural journals and has written many journal articles and technical reports.

CONTRIBUTORS

C. J. Andrews, Ph.D.
Research Scientist
Plant Research Centre
Agriculture Canada
Ottawa, Canada

Preston K. Andrews, Ph.D.
Irrigated Agriculture Research and Extension Center
Washington State University
Prosser, Washington

Edward N. Ashworth, Ph.D.
Plant Physiologist
Appalachian Fruit Research Station
USDA/ARS
Kearneysville, West Virginia

Tony H. H. Chen, Ph.D.
Research Scientist
Department of Horticulture
Oregon State University
Corvallis, Oregon

Leslie H. Fuchigami, Ph.D.
Professor
Department of Horticulture
Oregon State University

L. V. Gusta, Ph.D.
Senior Research Scientist
Crop Development Centre
University of Saskatchewan
Saskatoon, Saskatchewan, Canada

Renee M. Harber, B.Sc.
Graduate Research Assistant
Department of Horticulture
Oregon State University
Corvallis, Oregon

Norman P. A. Huner, Ph.D.
Associate Professor
Department of Plant Sciences
University of Western Ontario
London, Ontario, Canada

S. Iswari, Ph.D.
Research Associate
Department of Horticulture
University of Wisconsin
Madison, Wisconsin

Masaya Ishikawa, Ph.D.
Crop Development Centre
University of Saskatchewan
Saskatoon, Saskatchewan, Canada

Olavi Junttila, Dr.agric.
Professor
Institute of Biology and Geology
University of Tromsø
Tromsø, Norway

Alina Kacperska, Ph.D.
Professor
Faculty of Biology
University of Warsaw
Warsaw, Poland

Åse Kaurin, Dr.Sc.
Professor
Department of Biology and Geology
University of Tromsø
Tromsø, Norway

Marianna Krol, Ph.D.
Plant Physiologist
Department of Plant Sciences
University of Western Ontario
London, Ontario, Canada

Z. Krupa, Ph.D.
Assistant Professor
Department of Plant Physiology
Marie Curie Sklodowska University
Lublin, Poland

Paul H. Li, Ph.D.
Professor
Laboratory of Plant Hardiness
Department of Horticultural Science and Landscape Architecture
University of Minnesota
St. Paul, Minnesota

Ellen Maissan, M.Sc.
Plant Physiologist
Department of Botany
University of Toronto
Toronto, Ontario, Canada

Anne Swindeman Murphey, M.Sc.
Agricultural Research Technologist
Department of Horticulture and Landscape Architecture
Washington State University
Prosser, Washington

Jiwan P. Palta, Ph.D.
Associate Professor
Department of Horticulture
University of Wisconsin
Madison, Wisconsin

M. Keith Pomeroy, Ph.D.
Research Scientist
Plant Research Centre
Agriculture Canada
Ottawa, Ontario, Canada

E. L. Proebsting, Ph.D.
Horticulturist
Deaprtment of Horticulture
Irrigated Agriculture Research and Extension Center
Prosser, Washington

C. Rajashekar, Ph.D.
Department of Horticulture
Kansas State University
Manhattan, Kansas

Martin J. T. Reaney, Ph.D.
Crop Development Centre
University of Saskatchewan
Saskatoon, Saskatchewan, Canada

Guy F. Reisenauer, B.Sc.
Scientific Programmer
Washington State University
Prosser, Washington

Albert Robertson, Ph.D.
Crop Development Centre
University of Saskatchewan
Saskatoon, Saskatchewan, Canada

Linda Chalker-Scott, M.Sc.
Doctoral Candidate
Department of Horticulture
Oregon State Unviersity
Corvallis, Oregon

Matsuo Uemura, Dr.Sc.
Post Doctoral Fellow
Department of Agronomy
Cornell University
Ithaca, New York

Chien Yi Wang, Ph.D.
Research Horticulturist
Horticultural Crops Quality Laboratory
Beltsville, Maryland

John P. Williams, Ph.D.
Professor
Department of Botany
University of Toronto
Toronto, Ontario, Canada

Shizuo Yoshida, Dr.Sc.
Professor
Institute of Low Temperature Science
Hokkaido University
Sapporo, Japan

TABLE OF CONTENTS

PART I. RESPONSES OF CHILLING-INSENSITIVE PLANTS TO LOW STRESS TEMPERATURES

Chapter 1
The Induction of Cold Acclimation: The Role of Abscisic Acid 3
M. J. T. Reaney, M. Ishikawa, A. Robertson, and L. V. Gusta

Chapter 2
Metabolic Consequences of Low Temperature Stress in Chilling-Insensitive Plants 27
A. Kacperska

Chapter 3
Alterations of Plasma Membranes Related to Cold Acclimation of Plants 41
S. Yoshida and M. Uemura

Chapter 4
Development at Cold-Hardening Temperatures: Membrane Assembly and Organization 53
N. Huner, M. Krol, J. P. Williams, E. Malssan, and Z. Krupa

Chapter 5
The Role of Phenolic Compounds in Plant Stress Responses 67
L. Chalker-Scott and L. H. Fuchigami

Chapter 6
Ethylene-Induced Stress Resistance 81
R. M. Harber and L. H. Fuchigami

Chapter 7
Environmental Control of Growth Behavior and Cold Hardiness in Arctic and Subarctic Plants 91
O. Junttila and Å. Kaurin

Chapter 8
Low Temperature Injury in Winter Cereals 107
M. K. Pomeroy and C. J. Andrews

Chapter 9
Plasma Membrane H^+-ATPase as a Site of Functional Alteration during Cold Acclimation and Freezing Injury 123
S. Iswari and J. P. Palta

Chapter 10
Cryopreservation of Plant Cells and Organs 139
T. H. H. Chen and P. H. Li

Chapter 11
Properties of Peach Flower Buds which Facilitate Supercooling 153
E. N. Ashworth

Chapter 12
Forcasting Hardiness for Winter Cold Protection in Peach Orchards 159
E. Proebsting, P. Andrews, A. Swindeman-Murphey, and G. Reisenaver

PART II: RESPONSES OF CHILLING-SENSITIVE PLANTS TO LOW TEMPERATURE STRESS

Chapter 13
Mefluidide: A Synthetic Chemical that Protects Corn and Rice Seedlings from Chilling Injury 167
P. H. Li

Chapter 14
Relation of Chilling Stress to Ethylene Production 177
C. Y. Wang

Index 193

Responses of Chilling-Insensitive Plants to Low Temperature Stress

Chapter 1

THE INDUCTION OF COLD ACCLIMATION: THE ROLE OF ABSCISIC ACID

Martin J. T. Reaney, Masaya Ishikawa, Albert J. Robertson, and L. V. Gusta

TABLE OF CONTENTS

I. Introduction 4

II. Role of Growth Regulators in Cold Acclimation 4
A. Evidence for Translocatable Hardiness Promoters and Inhibitors 4
B. Endogenous ABA 5
C. Exogenous ABA 5
D. Gibberellins 6
E. Auxins 6
F. Cytokinins 7
G. Growth Regulators 7

III. Cross Adaptation 7

IV. ABA Responses 8

V. The Induction of Freezing Tolerance in Bromegrass Suspension Cultures by ABA 9
A. Factors Affecting ABA Induced Cold Hardening 9
1. Inoculum Weight 9
2. The Effect of ABA Concentration 9
3. Incubation Temperature 9
4. Ethanol 11
5. Miscellaneous Factors 11
6. Gibberellic Acid Effects on ABA-Induced Freezing Tolerance 12
7. Dehardening of ABA-Induced Cultures 13
8. The Effect of Sucrose on the ABA-Induced Cold Hardening 13
B. ABA-Induced Tolerance to Salt, Heat, and Osmotic Stresses 14
C. Morphological Cell Changes Induced by ABA 16
D. ABA and Low Temperature Induced Proteins 16

VI. Concluding Remarks 21

References 21

I. INTRODUCTION

The induction of freezing and chilling tolerance in temperate plant species requires a sensory mechanism for detecting environmental cues. The cue may be a stress or an environmental change, e.g., photoperiod. In response to the cues, metabolism switches from growth promotion to growth cessation. Short days at warm temperatures or low temperatures, or low temperatures alone are known to induce cold acclimation in temperate species.[1] Leaves perceive photoperiodic changes but the sensory mechanism is not known,[2] although phytochrome has been suggested to be involved.[3] There is considerable evidence that environmentally controlled, endogenous growth regulators such as abscisic acid (ABA) and gibberellic acid (GA) are involved in the adaptive response.

II. ROLE OF GROWTH REGULATORS IN COLD ACCLIMATION

A. Evidence for Translocatable Hardiness Promoters and Inhibitors

The leaves of woody plants produce both translocatable hardiness inhibitors[2,4] and translocatable hardiness promoters[2,4,5] depending upon the environmental cues. Irving and Lamphear[4,5] demonstrated that leaves of *Acer negundo* exposed to long days produced cold hardiness inhibitors which had GA-like activity, whereas extracts from leaves of plants exposed to short days had ABA-like activity.[4] Plants exposed to natural short days in the fall cold hardened, but plants receiving an extended photoperiod of 6 hr did not harden until the leaves were killed by a frost. Fuchigami[2] reported leaves of *Cornus sericea* growing at 15 to 20°C produced translocatable inhibitors of hardiness under long days and translocatable promoters of hardiness under short days. Removal of leaves from *C. sericea* exposed to long days enhanced acclimation, whereas removal of leaves from plants exposed to short days inhibited acclimation. A single defoliated branch of *C. sericea* hardened to the same level as the plant's remaining branches exposed to short days and low temperatures. However, the defoliated branch failed to acclimate when girdled at the base. In a continuing study,[2] two climatic clones of *C. sericea*, differing in time and rate of acclimation, were side grafted and pruned to two leaders. The grafted branches acclimated at the same rate as the parent clones. If the branch of the slow acclimating clone was defoliated, its acclimation rate increased, but it did not harden as rapidly or to the same extent as foliated branches from the fast acclimating clone. If the foliated, fast acclimating branch was girdled at the base, the rate and degree of acclimation of the defoliated, slow acclimating branch was reduced. In a reciprocal study, branches of the fast acclimating clone were defoliated, resulting in a decreased rate of acclimation. The above results suggest a translocatable hardiness promoter(s) is produced in leaves exposed to short days and translocatable hardiness inhibitor(s) is produced in leaves exposed to long days. The balance of the promoters and inhibitors may be more critical for hardening rather than the concentrations of the components involved.

Wheat leaves exposed to light at 2°C produced a cold hardiness promoter which was translocated to underground tillers of winter wheat (*Triticum aestivum*).[6] Plants exposed to light for as little as 15 min/day at 2°C hardened, whereas plants kept in total darkness did not. Tillers maintained in the dark hardened if the remainder of the plant was kept in the light at 2°C. In contrast, Shayakhmetova et al.[7] found that illuminated wheat leaves did not confer hardiness to leaves maintained in the dark. Limin and Fowler[8] reported that winter wheat leaves and crowns held at warm, nonhardening temperatures cold hardened if the roots were exposed to cool temperatures. However, if the leaves were maintained at low temperatures and the roots at warm temperatures, the leaves hardened but the crowns did not. Fayyaz et al.[9] found that a combination of short photoperiod and low air and soil temperature induced maximum cold hardiness in *Chrysanthemum morifolium* rhizomes. A long photoperiod, warm aerial temperatures, and low soil temperature did not induce rhizome hardening. Some hardening did occur in cool soils with either short photoperiods or low aerial temperatures. Therefore, the induction of rhizome hardening is dependent, in part, upon a stimulus translocated from the aerial to the below ground parts.

These translocatable hardiness promoters and inhibitors were not identified by the above authors. Several authors have suggested that ABA and GA may be the endogenous hardiness promoter and inhibitor, respectively.

B. Endogenous ABA

Irving and Lamphear[4,5] were among the first to suggest that ABA was the translocatable hardiness promoter. Higher levels of "dormin" or ABA were extracted from short-day induced *A. negundo* plants than from long day grown plants. Similarly, Mielke and Dennis[10] demonstrated an increase in ABA in *Prunus cerasus* L. in the autumn. When trees were mechanically defoliated prior to the onset of leaf abscission, no measurable increase in ABA was detected. The intracellular levels of ABA in the very hardy winter wheat cultivar Kharkov were threefold higher than in the less hardy cultivar Cappelle.[11] Exposure of winter rape seedlings to cold hardening conditions resulted in an increase in ABA.[12] However, Waldman et al.[13] found no difference in the ABA—like activity in fully hardened and nonhardened alfalfa (*Medicago sativa*). Daie et al.[14] measured an increase in ABA, hydrolyzable ABA and dihydrophaseic acid (DPA) in five warm season crops, but not in five cool season crops which were grown initially at 25°C and then exposed to 10°C. Chen and Li[15] reported a transitory threefold increase in ABA in the hardy potatoe species, *Solanum commersoni*, but not in the tender species *S. tuberosum* during the first 4 days exposure to hardening conditions. After 4 days, the level of ABA returned to its original level. In potato, ABA increased only in hardy species and only during the initial stages of hardening. This may explain why some researchers have not detected increases in endogenous ABA during hardening.

C. Exogenous ABA

Exogenous application of ABA to induce cold acclimation or chilling resistance has resulted in either no response, partial enhancement, or a dramatic increase. Irving and Lamphear[16] hardened twigs of *A. negundo* from $-15°$ to $-21.4°C$ by inserting leaves into vials containing a high concentration of ABA (100 mg/ℓ). Fuchigami et al.,[2] using a similar method, found no ABA enhancement of hardiness in *C. sericea*. Fayyaz et al.[9] could not induce hardening with ABA in *Chrysanthemum morifolium* rhizomes either at warm or cool temperatures. Gusta et al.[17] found that ABA inhibited growth but had little or no effect on enhancing freezing resistance of winter wheat crowns, when applied as a foliar spray or in the nutrient hydroponics solution to the roots. In contrast, Larson[18] increased the crown hardiness of Norstar winter wheat 12°C by an exogenous application of ABA to wheat seedlings at the two to three leaf stage. These plants were grown on agar under sterile conditions for 9 days in the presence of 7.5×10^{-5} *M* ABA. Plants exposed to 2°C hardened slightly faster in the presence of ABA compared to the controls, but the ultimate level of hardiness attained was similar. ABA was shown to prevent chilling injury in cotton (*Gossypium hirsutum*) and cucumber (*Cucumis sativus* L.) when added to plants in the light 18 hr prior to the chilling stress.[19] The light requirement could be replaced by soaking the tissue in a sucrose solution containing ABA.[19]

The most dramatic increases in freezing tolerance by exogenous ABA have been observed in potato leaf discs[15] and cell suspension cultures.[20] *S. commersonii* leaf discs increased 6°C when floated on ABA solutions of 25 mg/ℓ at 25°C.[15] Chen and Gusta[20] reported that cell suspension cultures of winter wheat, winter rye (*Secale cereale*), and bromegrass (*Bromus inermis*) treated with 7.5×10^{-5} *M* ABA for 4 days at 20°C could tolerate $-30°C$, whereas the control cultures tolerated only -7 to $-8°C$. Of ten cell-cultured species tested, ABA was effective in enhancing hardiness only of those species capable of cold hardening at low temperatures. ABA also induced similar responses in cell suspension cultures of alfalfa[21] and callus cultures of birdsfoot trefoil (*Lotus corniculatus* L.).[22]

There are several reasons why ABA may be more effective in inducing freezing resistance in tissue cultures than in intact plants. Tissue cultures are grown under sterile conditions in a defined media, and are exposed to a high level of ABA throughout the course of an experiment. In contrast, naturally grown plants are given temporary foliar applications of ABA which may be degraded by intense light[27] and microbes.[28]

D. Gibberellins

The relationship between GA and cold hardiness is complex. Most reports indicate a negative relationship between cold hardiness and endogenous GA levels.[2-4,16] Several researchers have shown that GA levels in plants decrease during hardening,[4,13,23] while others have found GA applications increased hardiness.[24,25] Waldman et al.[13] suggested that the ratio of GA to ABA controls the hardening process.

Irving and Lamphear[4,5] found higher levels of gibberellin-like compounds in long-day grown *A. negundo* than in short-day grown plants. Extracts from a hardy alfalfa cultivar had nearly no GA-like activity, whereas the GA activity did not change in a nonacclimating cultivar.[13] A significant decrease in GA activity was detected in fully hardened Kharkov winter wheat grown at 2°C as compared to plants grown at 20°C.[23] Andrews[26] found winter wheat seeds germinated in the light at 3.5°C were more hardy than seeds germinated in the dark; high levels of GA are produced in wheat plants germinated in the dark.[29] Lack of photosynthesis in the dark was not responsible for the decrease in cold hardening as the seedlings were still autotropic. In similar experiments, wheat seedlings grown at hardening temperatures required only a 15-min light exposure each day to achieve similar hardiness levels at plants grown in full light.[6] Phytochrome has been demonstrated to be involved in the hardening of winter rape (*Brassica napus*).[3] Red light or white light supplied for intervals as short as 3 hr were sufficient to inhibit dark induced etiolation and to increase plant cold hardiness, far red light and GA_3 counteracted the red light effect. These studies suggest that control of hardiness by light is under the control of phytochrome and GA levels rather than strict energy requirements.

Applications of GA to plants result in either decreased freezing tolerance, no effect, or a slight enhancement of freezing resistance. A foliar application of GA_3 (1000 mg/ℓ) reduced the freezing tolerance of short day grown *A. negundo* by 5°C.[4] Similarly, Fuchigami et al.[2] measured a 6°C reduction in the cold hardiness of *C. sericea* with foliar application of GA_3. Alfalfa seedlings sprayed with 2 m*M* GA every third day had elongated internodes[13] and reduced freezing tolerance,[30] even when grown under short days and low temperatures. The hardiness of Kharkov winter wheat was reduced when plants were grown in a nutrient solution containing GA_3 or GA_7.[23]

In contrast to these finding, other researchers have reported an enhancement of cold hardiness by exogenous application of GA.[24,25] However, the positive effect of GA was only observed when GA was applied during an active growth period. The differences observed due to GA may relate to the stage of plant development when GA was applied.

E. Auxins

Vasilyev[31] cites several studies which showed that auxins reduced cold hardiness. Although auxins are known to stimulate growth and cell division there are reports that auxins enhance cold hardiness. Holubowitz et al.[34] reported NAA sprays were beneficial for the development of cold hardiness and in subsequent cold injury repair. Sugawara and Sakai[33] observed that Jerusalem artichoke (*Helianthus tuberosus* L.) callus hardened slightly in response to 2,4-D at 26°C. Under hardening conditions, 2,4-D at an optimum concentration of 5×10^{-5} *M* stimulated the development of freezing tolerance. Poplar callus grown in a medium containing 3 mg/ℓ NAA were cold hardened at low temperatures to survive slow freezing to −196°C.[33] Chen and Gusta[20] hardened cultures by both low temperature and ABA treatment in media containing 1 mg/ℓ of 2,4-D. Therefore, the presence of auxins in the growth medium does not interfere with the hardening process.

F. Cytokinins

Cytokinins, a group of compounds responsible for cell division, were found to inhibit ABA-induced hardening of alfalfa cultures.[21] In contrast, kinetin did not inhibit the hardening of *S. acaule* leaf calli at 3°C in the dark.[15] Other reports have shown a positive effect of cytokinins on the cold hardiness of tomatoes (*Lycopersicon esculentum*),[37] barley (*Hordeum vulgare*), and peas (*Pisum sativum*).[38] Cytokinins applied to winter wheat grown in nutrient solution resulted in a twofold reduction in the crown water content, but little or no effect on hardiness was detectable except for a small increase when ABA was added.[17]

G. Growth Regulators

Grown cessation is considered a prerequisite for cold acclimation. Weiser[1] hypothesized cold hardiness was controlled by the balance of growth inhibitors and growth promoters. The cold hardiness of several species has been decreased by applications of GA and increased with growth retardants. This has led investigators to suggest that the balance of ABA and GA controls the hardening process. Several growth regulators e.g., cycocel (2-chloroethyl-trimethylaminosuccinamic acid), B-9 (succinic acid [2.2-dimethylhydrazide]) terpal, triazoles, etc. are known to inhibit GA biosynthesis. For a detailed description on growth retardants, the reader is referred to the review by Carter and Brenner.[39]

Kaperska-Palacz et al.[40] reported that cycocel and B-9 enhanced the hardiness of cabbage at low temperatures, but not at warm temperatures. Holubowitz et al.[34] found cycocel to be effective in enhancing acclimation only when applied as a foliar spray in late July. Gusta[41] applied several anti-GA compounds to winter wheat seedling in the fall. Although the growth regulations did not enhance the ultimate level of hardiness, treated plants retained their cold hardiness longer during the winter months than the controls.

Mefluidide increased the cold hardiness of *S. tuberosum* and induced chilling protection in *Zea mays* when applied 6 to 12 hr before stress induction.[43] Subsequently, Zhang et al.[44] discovered mefluidide increased endogenous ABA of corn plants grown in a nonstressful environment.

III. CROSS ADAPTATION

The development of tolerance to one stress may also increase the plant's tolerance to other stresses (Table 1). For example, a drought stress at nonhardening temperatures increased the frost tolerance of *Brassica oleracea*[45] in winter wheat,[46] and in *C. sericea*.[47] Cross adapation has also been noted in other combinations of stressful environments. Boussiba et al.[48] subjected tobacco (Nicotiana tabacum) plants to salinization, dehydration, mineral deficiency and boron toxicity. All four treatments decreased transpiration, increased freezing tolerance, and increased ABA levels more than twofold. Rikin and Richmond[49] found cucumber seedlings were more chilling tolerant after exposure to ABA, NaCl, and dehydration. Misrahi et al.[50] reported tobacco tissue had increased ABA levels when exposed to NaCl or a relative humidity of less than 90%. An induced nitrogen deficiency increased the winter hardiness of *Pinus sylvestra*,[51] *Festuca arundinaceae*,[52] *Forsythia intermedia*,[53] *Cynodon dactylon*,[54] and *Juniperus chinensis*.[55] Withholding nitrogen increases the level of ABA in plants.

Although a stress may confer some freezing resistance, the level of resistance does not

Table 1
SUMMARY OF RESEARCH SHOWING CROSS-ADAPTATION OF PLANTS TO STRESSES

Imposed stress	Induced tolerance	Species	Ref.
Drought	Freezing	Cabbage	45
		Dogwood	47
		Winter wheat	46, 56
		Tobacco	48
Salinity	Freezing	Tobacco	48
		Wheat	56
	Chilling	Cucumber	49
Nitrogen deficiency	Freezing	Pine	51
		Forsythia	53
		Festuca arundinaceae	52
		Cynodon dactylon	54
		Juniper	55
		Tobacco	48

approach that attained due to cold hardening conditions. For example, Tyler et al.[56] cold hardened winter wheat crowns to −9°C by exposing plants to drought and saline conditions at 20°C. Then cultivars can be cold hardened to −20°C by exposure to 2°C for 3 weeks.

IV. ABA RESPONSES

Exogenous application of ABA to plants affects the rate of growth, growth pattern, water flux, ion uptake, membrane fluidity, protein synthesis. ultrastructure, and cell differentiation.

Ballardin-Denti and Cocucci[57] reported 8×10^{-5} *M* ABA prevented membrane potential changes in germinating radish (*Raphanus sativus*) seeds and reduced the uptake of all major cations.[58] ABA accelerated ion uptake in beet roots (*Beta vulgaris*) and reversed the membranes normally selective uptake of K^+ in preference to Na^+.[59] Lado et al.[60] reported that ABA reduced H^+ extrusion in germinating radish seedlings. A similar decrease in proton efflux was seen in petioles of *Ricinus communis* floated on ABA solutions.[61] Potassium ion was exchanged for H^+ at the plasma membrane.

ABA alters water flux by changing the ion permeability and membrane potential,[62] for example, the rate of tritiated water efflux increased in ABA treated carrot (*Daucus carota*) discs.[63] An increase in root water permeability is reflected by an increase in root conductance as reported in soybean (*Glycine max*),[62] sunflower (*Helianthus annus* L.),[64] and corn.[65] Glinka[64] demonstrated water flow was stimulated within 30 min of ABA applications.

Tillberg et al.[66] demonstrated 10^{-7} *M* ABA inhibited the fresh weight increase in *Lemna gibba* by 30 to 60% as compared to the control; however, the relative dry matter content increased by 20 to 115% as compared to the control. Neskovic et al.[67] reported ABA at 3.8×10^{-6} *M* inhibited the growth of spinach (*Spinacea oleracea*) cultures, whereas at 3.8×10^{-8} *M*, growth increased compared to the control. Kochhar[68] found ABA accelerated tobacco callus growth (fresh weight and dry weight) at 3.5×10^{-6} *M*, but 3.5×10^{-5} *M* was inhibitory. Van Overbeek and Mason[69] observed ABA inhibited the fresh weight growth of *Lemna minor*; yet another study using the same species and concentration of ABA reported an increase in growth as determined from the dry weight of the tissue.[70]

GA_4 and GA_7 were synergistic in combination with ABA in promoting elongation of cucumber (*C. sativa*) hypocotyls.[71] High concentrations of synthetic ABA and natural ABA (10^{-4} *M*) were equally effective in enhancing the increase in fresh weight of soybean callus.[72]

Exogenous applications of ABA alters the solute composition and osmotic strength of tissues.[73] ABA stimulated proline accumulation in turgid barley leaves[74] but not in leaves of sunflower and tobacco (*N. tabacum*).[73] McLaren and Smith[70,75] reported a tenfold increase in sucrose and a twofold increase in fructose and glucose in *L. minor* cultures treated with 10^{-6} *M* ABA.

ABA stimulates the production of mRNA and proteins.[76] The activity of invertase increased in sugar cane (*Saccharum officinarum*) internodes treated with 10^{-4} to 10^{-7} *M* ABA,[77,78] Neskovic et al.[67] reported ABA (3.8×10^{-8} *M*) treatment of spinach (*S. oleracea*) callus resulted in an increase in RNA, DNA, and polysomes within 2 hr of treatment. In addition, spinach callus treated with 3.8×10^{-8} *M* ABA had smaller vacuoles and larger starch grains than controls. Various reports have detailed an increase in starch grain size with ABA treatment in a variety of tissues.[66,67,79,80] However, Chin and Weston[81] reported starch grain size was decreased in *L. esculentum* roots treated with 1.5×10^{-5} *M* ABA.

Cell differentiation is often restricted by ABA. Watson and Halperin[82] have shown the formation of xylem tube elements was reduced by ABA in cultures of jerusalem artichoke (*Helianthus tuberosum*). In spinach cultures, 3.8×10^{-8} *M* ABA increased the number of cells under 50 μm diameter, from 59 to 86% in 3 days.[67] Suberization in wounded potato tissues was prevented by continuous washing of the cut surface with distilled water.[83] When 10^{-4} *M* ABA was included in the washing solution suberization resumed.

Alfalfa seedlings grown in sterile agar containing 4×10^{-6} *M* ABA at 24°C with a 16-hr photoperiod attain a "rosette" or stunted pattern of growth, whereas the controls were elongated.[13] Plants grown under an 8-hr photoperiod at 5°C also displayed a "rosette" growth pattern, which is the typical growth habit of plants in the autumn.

V. THE INDUCTION OF FREEZING TOLERANCE IN BROMEGRASS SUSPENSION CULTURES BY ABA

The following sections summarize the research findings of the authors on ABA-induced freezing resistance of cell suspension cultures. Outlined are the factors which influence ABA-induced hardening and associated morphological and protein changes.

A. Factors Affecting ABA Induced Cold Hardening

1. Inoculum Weight

The weight of the cell inoculum, prior to the addition of ABA to the cultural medium, significantly affects the level of hardiness attained by the cells inn response to ABA (Table 2). An inoculum of 2 g or less treated with 7.5×10^{-5} *M* ABA for 7 days at 25°C had an LT_{50} of −36°C compared to an LT_{50} of −14°C for a 4 g cell inoculum. The 4 g cell inoculum had an LT_{50} of −19°C after 4 days of treatment with ABA; thereafter the cultures declined in hardiness due to a depletion of ABA, nutrients or other factor. In contrast the 2 g cell inoculum continued to harden from −7 to a LT_{50} of −36°C on day 7.

2. The Effect of ABA Concentration

Originally, Chen and Gusta[20] reported 7.5×10^{-5} *M* ABA as the optimum concentration for the induction of freezing tolerance in bromegrass cell cultures; the effect of ABA concentration on the induced hardening is shown in Figure 1. Due to changes in the original culture, a concentration of 4.5×10^{-5} *M* is now optimal for induction. Even at this high cincentration, the growth of ABA treated cells at 25°C was comparable to the growth of control cultures (data not shown).

3. Incubation Temperature

The optimum temperature for ABA induction of freezing tolerance of bromegrass cells

Table 2
THE EFFECT OF CELL INOCULUM SIZE ON THE FREEZING TOLERANCE OF CONTROL AND ABA TREATED BROMEGRASS CELLS

Inoculum Weight (g)	Days of Incubation	LT_{50} (°C) Control	LT_{50} (°C) ABA	LT_{50} (°C) (ABA — Control)
2	4	− 1.6	−19.4	−17.8
4	4	− 7.6	−18.6	−11.0
6	4	− 6.4	−18.4	−12.0
2	7	−10.4	−36.3	−25.9
4	7	− 9.5	−14.2	− 4.7

Note: Cells were incubated for 4 or 7 days in the dark at 25°C with or without ABA (7.5×10^{-5} *M*). The LT_{50} was estimated from TTC reduction tests.

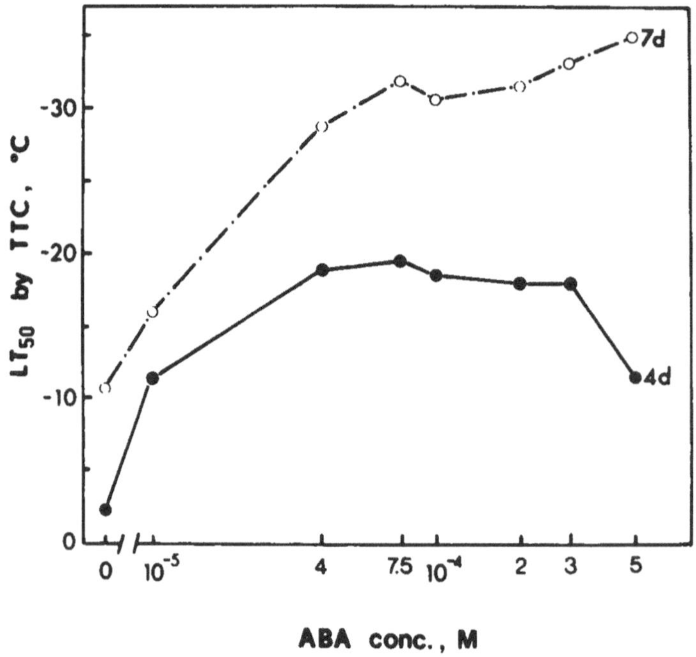

FIGURE 1. The effect of ABA concentration on ABA-induced cold hardiness in bromegrass cell suspension cultured cells. The cultures were inoculated with 2 g of cells and grown in the dark at 25°C up to 7 days. Cultures were grown for either 4 or 7 days as indicated. All freezing was initiated at −3°C followed by at least 1 hr equilibration time at 3°C. The LT_{50} was determined by TTC reduction after freezing the cells at 2°C/hr to −15°C and 5°C/hr to −40°C.

ranges from 25 to 30°C (Figure 2). Lower temperatures were less effective, resulting in only 12 to 18°C of increased freezing tolerance. Previous work established that 2 to 4°C was the most effective temperature for inducing freezing resistance in non-ABA treated cells.[97] Control cells were the least frost tolerant after 4 days of subculture at 20 to 30°C. After 7 days of subculture at 25 to 30°C, the cells hardened an additional 4°C, possibly in response

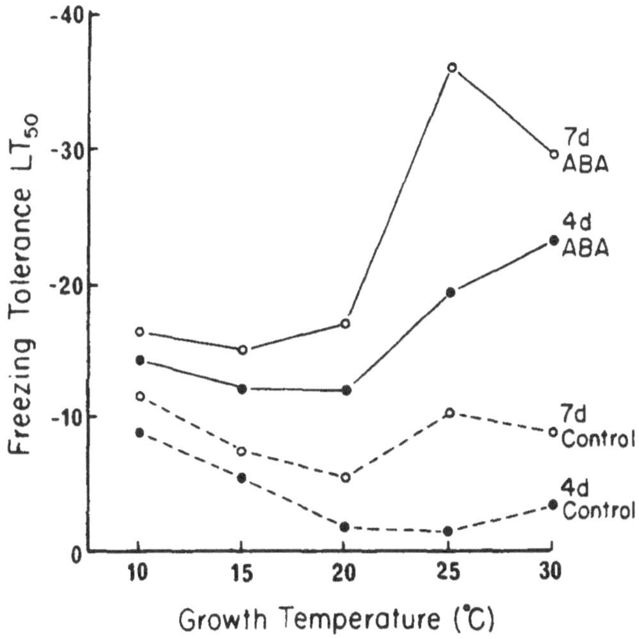

FIGURE 2. The effect of temperature on ABA-induced (7.5×10^{-5} *M*) cold hardiness in bromegrass cell suspension cultured cells. Growth conditions, freezing conditions and viability assays are described in Figure 1. Cultures were grown for either 4 or 7 days at the temperatures indicated.

to aging and nutrient depletion. Cells maintained at 15 to 20°C did not show this age-induced hardening.

The growth rate of cells treated with ABA was comparable to control cells at all temperatures tested. Cells maintained at 25 to 30°C in the presence of ABA grew the fastest and developed the most freezing tolerance (Figure 2), in direct contrast to low temperature-induced hardening in plants where growth is suppressed.[84] Reportedly, high concentrations of ABA generally suppress growth,[85] however, ABA has been shown to enhance embryo growth[86] and stimulate mesocotyle elongation.[87] The bromegrass culture used in this study originated from the mesocotyl tissue of seedlings, and this may account for the growth of cells in the presence of high levels of ABA.

4. Ethanol

Ethanol is often used to aid solvation of ABA, which is sparingly water soluble. Only small quantities of ethanol evaporate during autoclaving due to the colligative property of vapor pressure depression. Ethanol added to the medium after autoclaving, at concentrations higher than 5 m*M* dramatically inhibited the ABA induced cold hardiness (Table 3). Yet, ethanol did not affect the cold hardiness of the control cultures at 25°C. Concentrations higher than 25 m*M* inhibited the growth of control and ABA treated cells. ABA and other growth regulators are routinely dissolved in an alkaline solution, e.g., NaOH, prior to addition to the medium.

5. Miscellaneous Factors

Bromegrass cultures grown at 25°C for 7 days were 8 to 10°C more freezing tolerant than cultures grown for 3 to 4 days at 25°C (Figure 2). This age-induced hardening was prevented by supplementing the media on day four with nitrogen nutrients (data not shown). Similarly,

Table 3
THE EFFECT OF ETHANOL ON ABA-INDUCED COLD HARDENING

Ethanol Conc. (m*M*)	LT_{50} (°C)		Fresh weight increase	
	Control	ABA	Control	ABA
0	−8	−32	100	100
5	−10	−28	100	102
10	−9	−20	97	104
25	−10	−20	77	32
50	−9	−20	64	26

Note: The cells (2 g inoculum) were incubated at 25°C in the dark for 7 days. The media contained 0 to 50 m*M* ethanol which was added aseptically after autoclaving and ABA (7.5×10^{-5} *M*) as described.

Table 4
THE EFFECTS OF GIBBERELLINS (GA_3, GA_4, GA_7, AND GA_9) ON ABA INDUCED FREEZING TOLERANCE OF BROMEGRASS SUSPENSION CULTURE CELLS

	LT_{50}							
	GA_3		GA_4		GA_7		GA_9	
Concentration (*M*)	−ABA	+ABA	−ABA	+ABA	−ABA	+ABA	−ABA	+ABA
0	−6	−32	−6	−32	−6	−32	−6	−32
2×10^{-6}	—[a]	—	—	—	—	—	−8	−36
5×10^{-6}	—	—	—	—	—	—	−4	−16
2×10^{-5}	−6	−32	−6	−13	−6	−12	−4	−15
5×10^{-5}	−6	−32	−6	−14	−6	−14	−8	−13
2×10^{-4}	−7	−30	−8	−13	−8	−11	—	—
5×10^{-4}	−7	−29	0[b]	−13	0[b]	−6	—	—

Note: All conditions are identical to those described in Figure 2 grown at 25°C in the dark.

[a] Data not available.
[b] Lethal treatment.

in a study on the effect of NH_4^+, NO_3^-, $PO_4^{=}$ and K^+ on hardening, only NH_4 and NO_3^- partially reduced the ABA enhancement of freezing tolerance.

Cells harden to the same extent in the presence or absence of light. Precautions to prevent light-induced temperature effect are necessary as temperatures 5°C higher than ambient inside of culture flasks have been recorded.

6. Gibberellic Acid Effects on ABA-Induced Freezing Tolerance

As described earlier, exogenous applications of gibberellins reduce the cold hardiness of plants. The effect of GA_3, GA_4, GA_7, and GA_9 on the ABA-induced cold hardiness of bromerass was determined (Table 4). Ga_3 had little effect on the freezing tolerance of the control or ABA treated cells. Concentrations of 2×10^{-5} *M* and higher, however, GA_4, GA_7, and GA_9 reduced the freezing tolerance of ABA treated cells from an LT_{50} of −32 to −13°C.

Table 5
THE EFFECT OF EXOGENOUS SUCROSE (10% w/v) ON THE FREEZING RESISTANCE OF BROMEGRASS CELL SUSPENSION CULTURES

	LT_{50} (°C)	
	Control	ABA
4 days		
− sucrose	−2.8	−23.0
+ sucrose	−6.5	<−40*
7 days		
− sucrose	−10.8	−32.0
+ sucrose	−15.6	<−40*

Note: *100% survival after −40°C. Sucrose was added to the cells 1 hr prior to the freezing tests. The cells (2 g inoculum) were incubated at 25°C for 4 or 7 days in the presence or absence of ABA (7.5 × 10^{-5} *M*). The LT_{50} was determined by TTC reduction tests.

At a concentration of 4 × 10^{-4} *M*, a mixture of GA_4, GA_7, GA_9 completely inhibited the growth rate of cells and was lethal to the cells at 10^{-3} *M*. ABA overcame the lethal effect of GA_4 and GA_7 but did not restore the growth rate.

7. Dehardening of ABA-Induced Cultures

To determine if ABA is required for the maintenance of freezing tolerance, ABA cold-hardened cells (LT_{50} = −35°C) were transferred to a fresh medium without ABA. After 7 days at 25°C, the bromegrass cells had an LT_{50} of −18°C, although 20% of the cells had an LT_{50} greater than −30°C. These results suggest that ABA must be present in the medium at warm temperatures to maintain a high level of freezing tolerance and that cells do not lose cold hardiness uniformly.

8. The Effect of Sucrose On the ABA Induced Cold Hardening

The effect of 10% sucrose on the freezing tolerance of bromegrass cells after treatment with 7.5 × 10^{-5} *M* ABA for 4 and 7 days was determined (Table 5). Cells treated with ABA for either 4 or 7 days and exposed to 10% sucrose 1 hr prior to freezing had a LT_{50} greater than −40°C. ABA treated cells not exposed to 10% sucrose had LT_{50}s of −23 and −32°C after 4 and 7 days, respectively (Table 5). Control cells exposed to 10% sucrose were 4 to 5°C more freezing tolerant than untreated control cells. It is unlikely that was taken up by the cells within the 30-min period. Sucrose is a well-known cryoprotectant, although its mode of action is not understood. Sucrose may influence the hardiness of cells by binding to the plasma membrane to prevent dehydration,[101] by partially plasmolyzing the cells and reducing cell wall effects.[10]'s2 Gusta[41] has demonstrated via nuclear magnetic resonance (NMR) spectroscoopy that cells with rigid walls lose water more slowly during the initial stages of freezing than cells with flexible walls. Rapid loss of intracellular water is necessary to prevent lethal intracellular freezing.

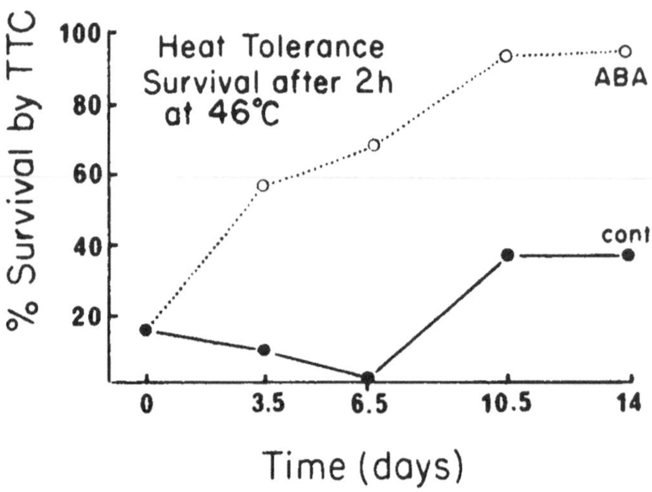

A

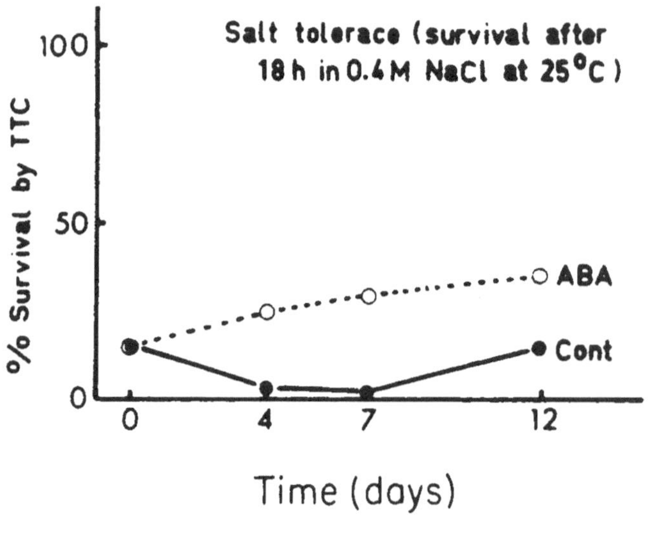

B

FIGURE 3. Induction of heat and salt tolerance by ageing and ABA (7.5 $\times$ 10^{-5} *M*) in bromegrass cell suspension cultures. Growth conditions are described in Figure 1. (A) Survival of 46°C for 2 hr in the dark, estimated by TTC reduction test. (B) Survival after incubation in 0.4 *M* NaCl in the dark at 25°C for 18 hr, estimated by TTC reduction tests.

B. ABA-Induced Tolerance to Salt, Heat, and Osmotic Stresses

Bromegrass cells were used as a model system to determine whether ABA induces tolerance to heat and salt stresses (Figures 3, 4). A 1 g inoculum, was grown in the presence of 7.5 $\times$ 10^{-5} *M* ABA for 14 days. The survival of cells exposed to 46°C for 2 hr was determined by TTC (Figure 3a). Within 84 hr, the heat tolerance of ABA-treated cells was significantly greater than the controls which were very heat susceptible. Maximum heat tolerance of ABA-treated cells occurred after 10.5 days. The heat tolerance of the control cells also increased after 10.5 days of subculture, but not to the same extent as the ABA-treated cells. This may be related to aging of the control cultures as was previously shown for cold

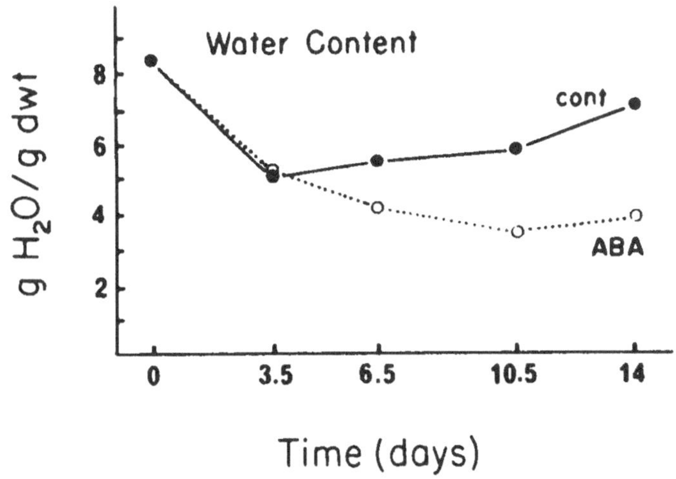

A

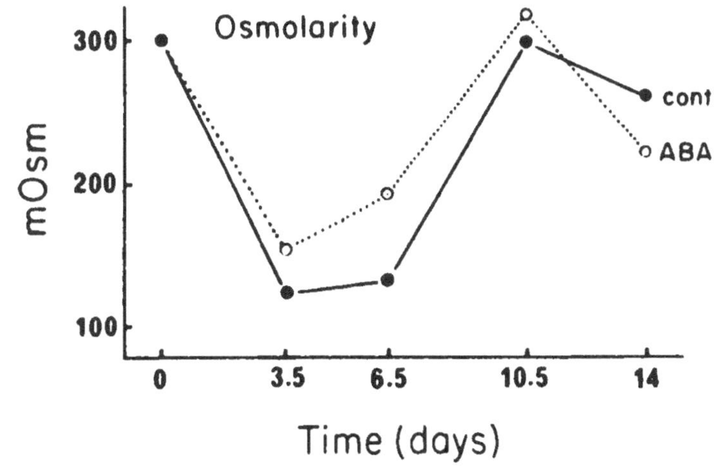

B

FIGURE 4. The effect of ABA on relative water content and osmotic strength of bromegrass cell suspension cultures. Growth conditions are described in Figure 1. (A) The relative water content was determined on tissue culture after removing extracellular water at 200 gravities for 5 min. Dry weights were determined after drying at 70°C for 48 hr. (B) Osmolarity of expressed cell sap was determined by a Wescor vapour pressure osmometer.

hardening. This increase in stress tolerance was prevented by the addition of NH_4^+ or NO_3^- after 4 days (results not shown).

Salt tolerance was determined by TTC after ABA-treated cells were subjected to a 0.4 *M* NaCl medium solution at 25°C for 18 hr (Figure 3B). Although the increase in salt tolerance for ABA-treated cells were not as dramatic as observed for the freezing and heat stresses, ABA-treated cells were more salt tolerant than the controls after 4 days.

Several physiological parameters were determined during the culture period. The water content of cells decreased from 8.5 to 5 g of H_2O/g dry wt within 3.5 days in both the control and ABA-treated cells (Figure 4A). Thereafter, the water content of control cells increased slightly to a final concentration of 7 g H_2O/g dry weight on day 14. The water content of the ABA-treated cells decreased to 4 g H_2O/g dry weight after 10.5 days and remained constant thereafter. The osmolality of the cell sap of control and ABA-treated cells dropped from an initial level of 300 to 125 mOsm for control cells and to 160 mOsm for the ABA-treated cells (Figure 4B). On day 6.5, the osmolality of the ABA-treated cells increased to 200 mOsm while the osmolality of the controls remained constant at 125 mOsm. Maximum osmolality of the cell sap occurred on day 10.5 reaching a concentration of 300 mOsm for both the control and ABA-treated cells. Therefore, these results suggest that osmotic adjustment alone does not account for the increase in ABA-induced stress tolerance.

C. Morphological Cell Changes Induced by ABA

After 7 days incubation at 25°C with 7.5×10^{-5} *M* ABA, cells were typical of the augmented hardy cells described by Siminovitch et al.[88] The large central vacuole disappeared to form small vacuoles, and lipid bodies and starch granules increased. The cell walls of ABA-treated cells thickened dramatically. After 2 weeks of growth in the presence of ABA, the intracellular structure was not readily discernible by light microscopy due to the thick cell walls. Histochemical tests suggested an increase in cell wall phenolics, possibly lignins. The ABA treated cells were more difficult to digest enzymatically than control cells, although it was still possible to produce protoplasts.

D. ABA and Low Temperature Induced Proteins

It is well established that protein changes occur during the development of cold hardiness.[89] Cycloheximide, an inhibitor of cytoplasmic protein synthesis, blocks the developement of cold hardiness.[90] Numerous electrophoretic analyses show an increased synthesis of soluble proteins during cold acclimation. Analyses of plasma membrane fractions extracted from winter rye seedlings[91] and orchard grass[92] show qualitative changes in the protein complement during acclimation. Recently, Guy et al.[93] identified mRNA transcripts which coded for two high molecular weight polypeptides which appear during the development of freezing tolerance in spinach.

Often, a specific correlation cannot be made between the protein changes identified and freezing resistance due to other growth processes. Processes such as vernalization, dormancy, adaptation to low temperature growth, and developmental changes occur simultaneously with cold acclimation. Enhanced synthesis of several proteins have been catalogued after ABA treatment of barley aleurone layers[94] and wheat embryos.[95] In addition, Heikkila et al.[76] demonstrated that ABA enhanced the synthesis of a 70 kdalton protein, which was also induced by heat shock and water stress.

ABA treatment and low temperature induce freezing resistance in suspension cultures with the genetic potential to cold acclimate. Therefore, proteins induced by low temperatures and ABA treatment are possibly involved in the developmental and functional aspects of freezing resistance. Conversely, proteins induced by ABA and low temperature treatment alone may not be associated with freezing resistance, but are associated with ABA metabolism and aspects of low temperature growth.

To study protein synthesis during hardening, low temperature and ABA-treated bromegrass cultures were equilibrium labeled with [^{14}C]-leucine. Total cell protein fractions from [^{14}C]-leucine labeled cells were separated by one dimensional SDS PAGE. The fluorogram in Figure 5 (8hr, cf arrows) indicates that within 8 hr ABA treatment at 23°C induced the synthesis of two low molecular weight proteins (19 and 21 kdaltons), and inhibited the synthesis of a 66 kdalton protein seen in 23°C controls. Low levels of label were incorporated in cultures grown at 3°C with or without ABA after 8 hr of incubation.

After 36 hr at 3°C, significant levels of [^{14}C]-leucine were incorporated in both treatments, and ABA inhibited the synthesis of the 66 kdalton polypeptide band seen at 23°C. After 36

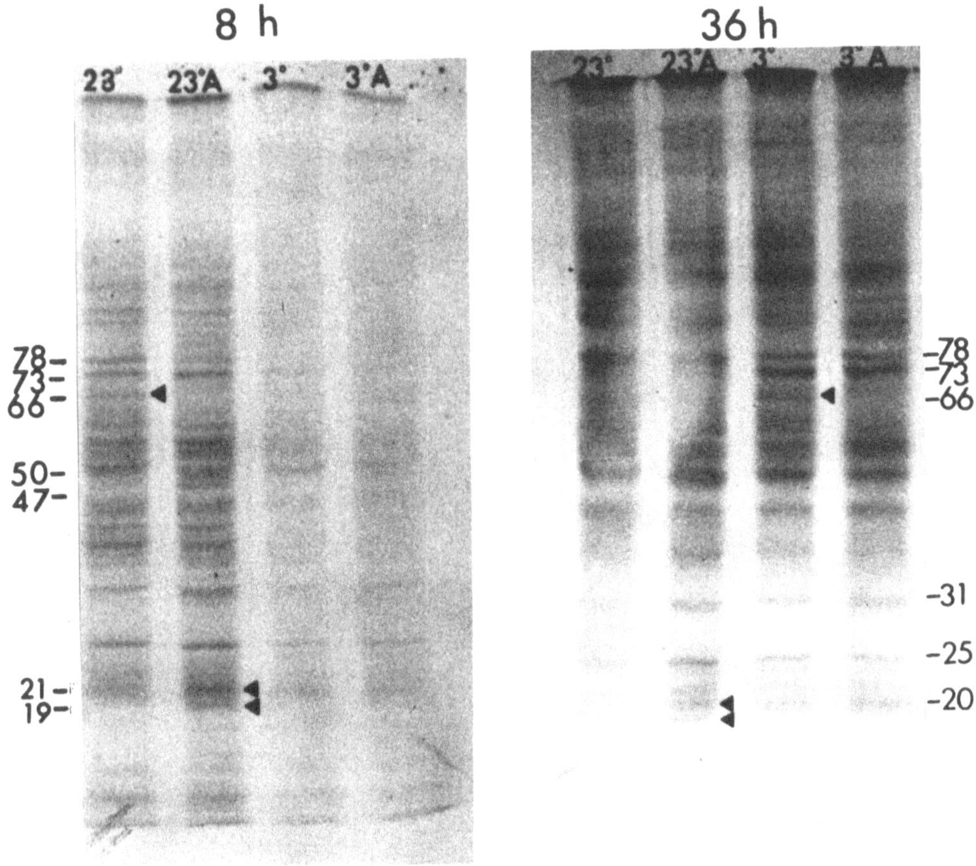

FIGURE 5. Flourograms of [^{14}C]-leucine labeled bromegrass cell protein extracts obtained after 8 and 36 hr of growth in the presence (23°C ABA, 3°C ABA) and absence (23°C, 3°C) of ABA. Each lane contained 60 μg of total cell protein. Dried gels were exposed to X-ray film for 20 days at −80°C.

hr at 23°C, considerable degradation and turnover of polypeptides occurred, however, in the presence of 7.5×10^{-5} *M* ABA, bromegrass cells either accumulated larger amounts or more stable forms of 31, 25, and 20 kdalton proteins (Figure 5). These same polypeptides are major proteins of low temperature grown cultures. This implies that the 3°C incubation conditions may result in synthesis of proteins similar to those induced by exogenous ABA treatment at 23°C.

After 86 hr, cells induced to cold harden with 7.5×10^{-5} *M* ABA at 23°C had an LT_{50} of −24°C vs. −8°C for control cultures. The LT_{50} values of cells hardened at 3°C with and without 7.5×10^{-5} *M* ABA were −14 and −16°C, respectively. Total cell protein fractions from the above treatments were analyzed after 86 hr of incubation by two-dimensional electrophoresis. At this time, the 3°C, 3°C plus ABA, and 23°C plus ABA-treated cells had high levels of freezing tolerance. Coomassie blue staining of the two-dimensional gels indicated an enrichment of a major polypeptide with an apparent M_r of 31.5 kdalton and an apparent PI of 6.85 in cells cultured at 23°C in the presence of ABA. The 31.5 kdalton protein was also enriched in 3°C and 3°C plus ABA protein separations (data not shown). Figure 6 shows the fluorograms obtained after separation of proteins from 23°C, 23°C plus ABA, 3°C, and 3°C plus ABA cells incubated for 86 hr in the presence of [^{14}C]-leucine. The higher turnover and degradation rates that occur at 23°C as compared to 3°C is reflected in the fluorograms. A 31.5 and 72 kdalton polypeptide is enriched in ABA treated cultures

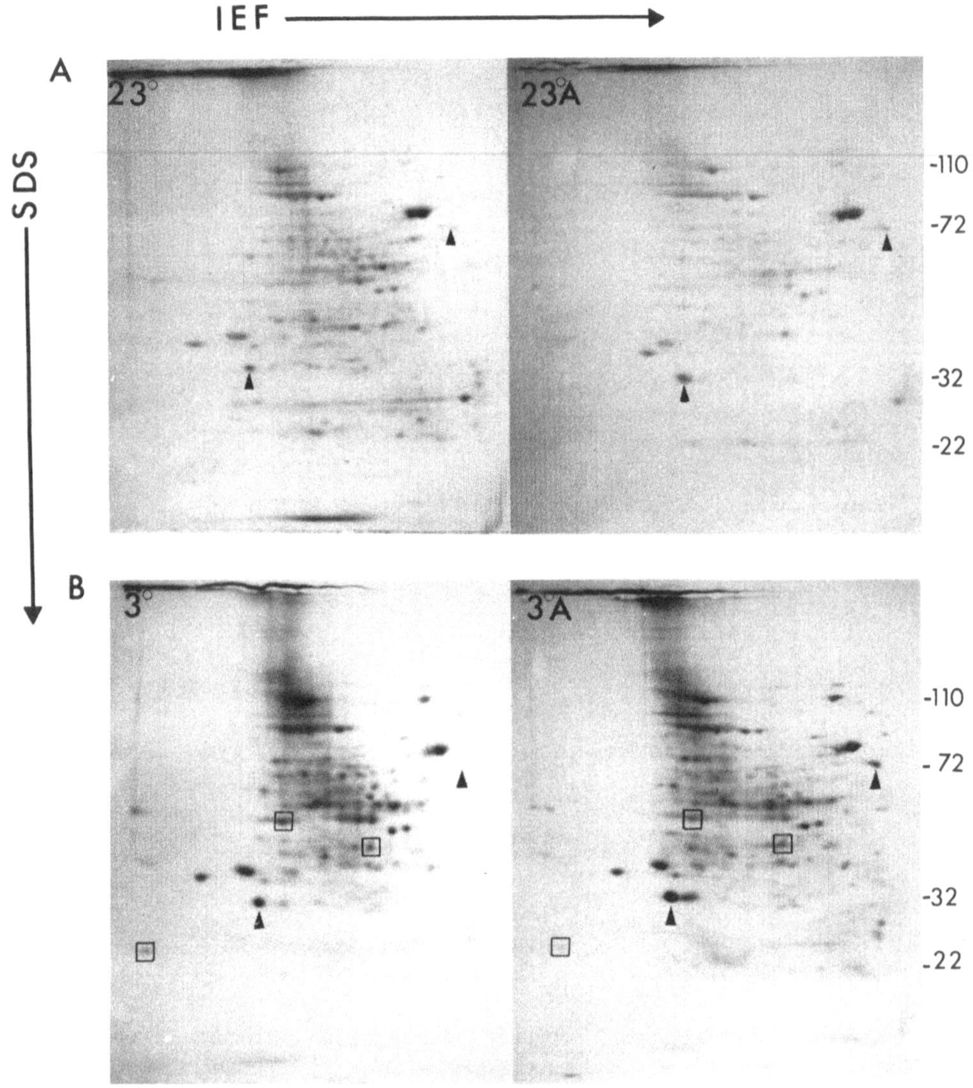

FIGURE 6. Two-dimensional flourogram of total cell protein extracts from cultured bromegrass cells incubated for 86 hr in media containing [^{14}C]-leucine. In Figure 2A cultures were incubated in the presence (23° ABA) or absence (23°) of 75 *M* ABA. Figure 2B shows the results of the low temperature treatment (3°C and 3°C ABA). The pH gradient was from pH 7.5 (left) to 5.5 (right). Arrow depicts polypeptides (31.5 and 72 kdaltons) enriched in ABA treated cultures at both 23°C and 3°C incubation temperatures. Three polypeptides (24.5, 48, 47 kdaltons) were observed only at 3°C growth temperature and are enclosed in squares.

at both 23 and 3°C (Figure 7, cf. arrowheads). The 31.5 kdalton protein has an apparent M_r similar to the 32 kdalton protein shown to be enriched by one-dimensional analysis of 23°C ABA, 3°C, and 3°C ABA cultures after 36 hr of incubation (Figure 5). The 20 and 25 kdalton proteins enriched in one-dimensional analysis of [^{14}C]-leucine labeled cultures after 36 hr of incubation were not apparent in the two-dimensional separations shown in Figure 6. These proteins have basic isoelectric points and either focus outside of the pH gradient analyzed here or become rapidly metabolized and are not detectable after 86 hr incubation periods. Three polypeptides were observed only at 3°C growth temperatures (24.5, 47, and 48 kdaltons). It is possible that the 24.5, 47, and 48 kdalton proteins are turned over at 23°C, or are not apparent in protein separation of 23°C cells, or are unique to low temperature growth.

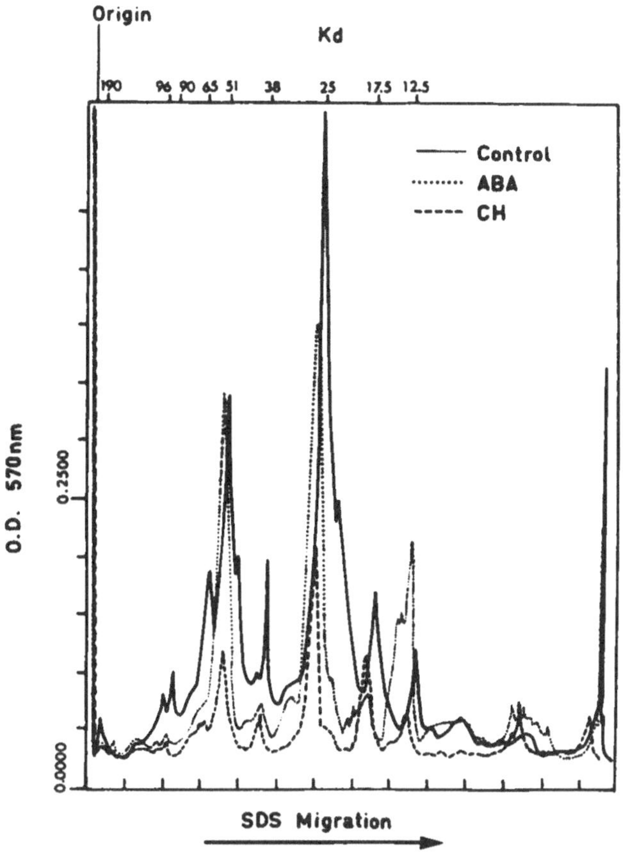

FIGURE 7. Densitometer tracings of Coomassie blue stained protein patterns obtained from extracellular filtrates of control (C), ABA (A), and low temperature (CH) treated alfalfa suspension cultures. Major protein bands showed decreased electrophoretic mobilities in ABA and low temperature hardened cultures, with apparent molecular weight values of 2 to 3 kdaltons higher. Exogenous ABA treatment induced the synthesis of 12.5 and 13 to 15 kdalton proteins.

In total cell protein fractions from ABA and low temperature hardened alfalfa cells, an enrichment of 42 kdalton protein was found. In protein studies of both alfalfa and bromegrass suspension cultures, a general decrease in the mobility (1 to 3 kdaltons) of proteins isolated from cold hardened and ABA-treated cells was observed. Similar changes in the electrophoretic mobility of major polypeprides and glycopeptides during natural and artificial cold hardening was observed in protein patterns of mulberry cortex,[96] winter rye,[97] and jerusalem artichoke tubers.[98]

Cellular fractionation studies were performed to localize ABA and low temperature induced proteins. Extracellular and subcellular fractions were collected from both alfalfa and bromegrass suspension cultures. Figure 7 is a tracing obtained from a Coomassie blue stained electrophoretic profile of extracellular proteins collected from 4- to 6-day-old control, ABA (LT_{50} = $-12°C$) and low temperature (LT_{50} = $-13°C$) grown alfalfa cells. Proteins from both ABA and low temperature grown alfalfa cells showed decreased electrophoretic mobilities. Exogenous ABA treatment resulted in enhanced synthesis or the induction of low molecular weight proteins of 12.5 and 13 to 15 kdaltons (Figure 7). The ABA-induced proteins were not enhanced in the low temperature grown cultures. A similar analysis was

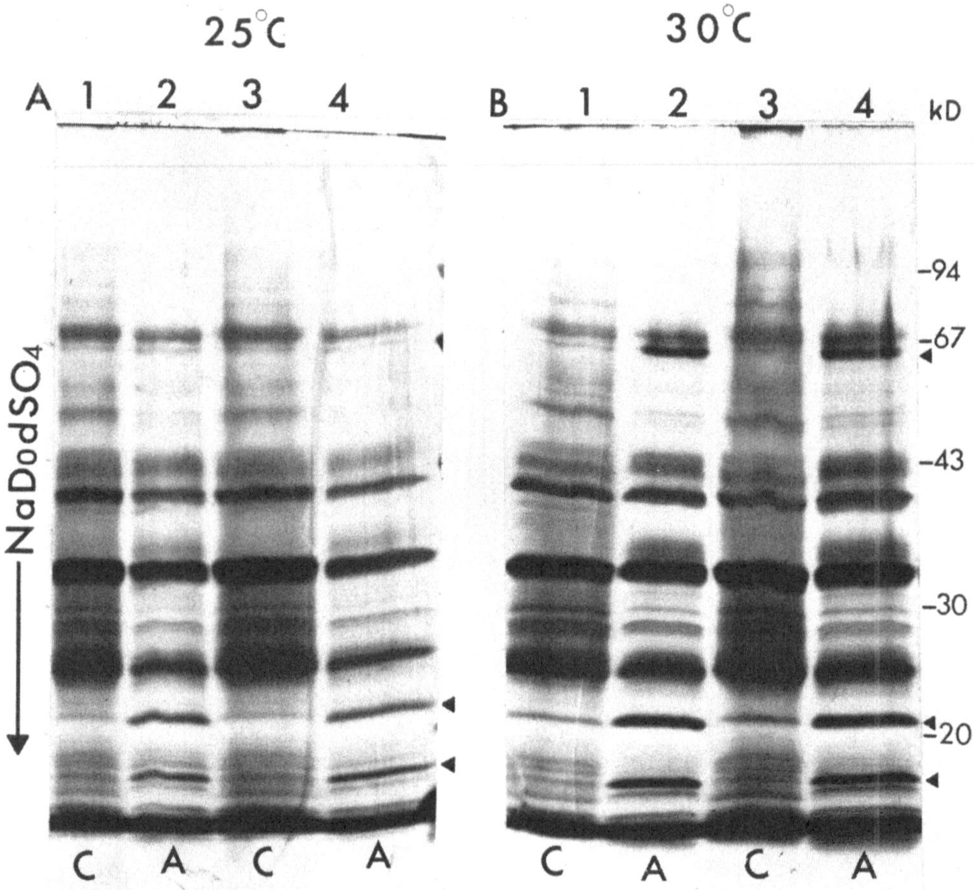

FIGURE 8. Silver stained sodium dodecyl sulfate polyacrylamide gels of extracellular protein fractions from bromegrass cell cultures. (A) Protein profiles of extracellular protein from control (C) and cultures grown in the presence of 7.5 *M* ABA (A) at 25°C for 7 days. Lanes 1 and 2 contain 50 g of protein and lanes 3 and 4, 100 g of protein. (B) as in A except cultures were incubated at 30°C. Arrows indicate proteins induced by ABA treatment at 25°C (17 and 21 kdaltons) and at 30°C (17, 21, and 65 kdaltons).

conducted on extracellular proteins from control (LT_{50} = −10.2°C) and 7.5 × 10^{-5} *M* ABA treated (LT_{50} = −36.2°C) bromegrass cells incubated at 25 and 30°C (Figure 8) was performed. Exogenous ABA treatment resulted in the induction of 21 and 17 kdalton proteins which were clearly detected in culture filtrates. Incubation of bromegrass cells at 30°C in the presence of ABA resulted not only in the expression of a 65 kdalton protein but also in a 6° decrase in LT_{50}. Sephadex gel column chromatography and amino acid analysis of the extracellular fractio suggest that the ABA-induced proteins (21 and 17 kdaltons) are rich in arginine and lysine amino acids. These proteins focus in the basic region of a nonequilibrium pH gradient.

ABA-treated alfalfa cultures showed an enrichment in a 24 kdalton protein on cell-wall fractions however, no enrichment of the 24 kdalton protein was seen in alfalfa cultures hardened by low temperature treatment. A similar analysis of soluble protein fractions from ABA-hardened alfalfa cultures indicated increased levels of three polypeptides with apparent molecular weights of 41, 37, and 20 kdaltons. The 42 kdalton protein in low temperature grown and ABA-treated cultures may be involved in the freezing resistance in alfalfa cultures. In bromegrass cultures hardened by ABA and low temperatures, a 20.2 kdalton protein was enriched in the soluble fraction and a 29 kdalton protein cofractionated with the crude membrane fractions. Two-dimensional analysis of the crude membrane fractions suggests that the 31.5 kdalton protein, previously identified in fluorograms of 23°C ABA, 3°C, and 3°C ABA fractions, is a membrane protein. This polypeptide was enriched in crude membrane fractions isolated from both cold hardened and ABA treated cultures after 4 to 6 days of culture. It is tempting to speculate that the 31.5 kdalton protein may be involved in the maintenance of freezing resistance in bromegrass cultures and the 42 kdalton polypeptide may be involved in the maintenance of freezing resistance of alfalfa cells. Polypeptides induced by ABA at earleier time periods (19 and 20 kdaltons) may be involved in triggering the cold hardening response.

VI. CONCLUDING REMARKS

There are several mechanisms by which plant growth regulators control cold hardiness. Gene transcription and protein synthesis are required for the manifold changes involved in hardening. Genetic constituents determine the maximum levels of hardening for any plant, but regulatory genes control the maximum level of hardiness attained.

The overall importance of regulation is demonstrated by the effectiveness of plant growth regulators in enhancing plant cold hardiness. The mode of action of plant growth regulators may be the inhibition of GA biosynthesis, but this has not been established to be the only mode of action. GA synthesis and inactivation is a complex process involving several biologically active molecules. Furthermore, It has been established that ABA breakdown products phaseic acid and dihydrophaseic acid also possess biological activity in some systems. Other endogenous plant growth regulators are also associated with cold hardiness. Therefore, the interaction of all these molecules may be important to cold hardiness.

Plants with superior regulation of cold hardiness may be found in breeding programs by current selection practices. There is a strong positive correlation ($r^2 = 0.05$) between prostrate growth habit and cold hardiness in winter wheat.[99] Also, there is a negative correlation between winter barley mesocotyl elongation and cold hardiness.[100] In these cases the underlying regulatory functions of cold hardening are expressed by their effects on growth habit.

Plants possessing identical genetic backgrounds may not have the same level of cold hardiness. For example, winter rye (Puma) as a diploid tolerates −28°C (LT_{50}) whereas the tetraploid tolerates −23°C.[101] These differences may be in part due to gene regulation.

The objectives of this study of growth regulator effects on cold hardiness may be to separate regulator from other gene effects. In this way the level of hardiness attainable with the optimum gene regulation can be established. Furthermore, it can be determined which proteins are regulated and how each of these proteins affect cold hardiness.

REFERENCES

1. **Weiser, C. J.,** Cold resistance and injury in woody plants, *Science*, 169, 1269, 1970.
2. **Fuchigami, L. H.,** The Environmental Physiological and Chemical Control of Cold Acclimation in *Cornus stolonifera* Michx., Ph.D. thesis, University of Minnesota, St. Paul, 1970.
3. **Kacperska-Palacz, A., Debska, Z., and Jakubowska, A.,** The phytochrome involvement in the frost hardening process of rape seedlings, *Bot Gaz. (Chicago)*, 136, 137, 1975.
4. **Irving, R. M. and Lanphear, F. O.,** The long day leaf as a source of cold hardiness inhibitors. *Plant Physiol.*, 42, 1384, 1967.
5. **Irving, R. M. and Lanphear, F. O.,** Environmental control of cold hardiness in woody plants, *Plant Physiol.*, 42, 1191, 1967.
6. **Trunova, T. I., Smirnova, N. A., and Zvereva, G. N.,** Role of light in development of frost resistance of plants, *Sov. Plant Physiol.*, 5, 109, 1976.

7. **Shayakhmetova, I. S., Trunova, T. I., and Mashchanova, D. D.,** The role of light in frost resistance development in winter cereals, *Physiol. Plants,* 30, 1208, 1983.
8. **Limin, A. E. and Fowler, D. B.,** Cold-hardiness response of sequential winter wheat segments to differing temperature regimes, *Crop Sci.,* 25, 838, 1985.
9. **Fayyaz, M. M., McCown, B. H., and Beck, G. E.,** Effect of temperature, photoperiod and several growth substances on the cold hardiness of *Chrysanthemum morifolium* rhizomes, *Physiol. Plant,* 44, 73, 1978.
10. **Mielke, E. A. and Dennis, F. G.,** Hormonal control of flower bud dormancy in sour cherry (*Prunus cerasus* L.). III. Effects of leaves, defoliation, and temperature on levels of abscisic acid in flower primordia, *J. Am. Soc. Hortic. Sci.,* 103, 446, 1978.
11. **Wightman, F.,** Modern chromatographic methods for the identification and quantification of plant growth regulators and their application to studies of the changes in hormonal substances in winter wheat during acclimation to cold stress conditions, in *Plant Regulation and World Agriculture,* Scott, T. K., Ed., Plenum Press, New York, 1979, 327.
12. **Kacperska-Palacz, A.,** Mechanism of cold acclimation in herbaceous plants, in *Plant Cold Hardiness and Freezing Stress,* Li, P. H. and Sakai, A., Eds., Academic Press, New York, 139, 1978.
13. **Waldman, M., Rikin, A., and Richmond, A. E.,** Hormonal regulators of morphogenesis and cold resistance.II. Effect of cold-acclimation and of exogenous abscisic acid on gibberelic acid and abscisic acid activities in alfalfa *Medicago sativa* seedlings, *J. Exp. Bot.,* 26, 853, 1975.
14. **Daie, J., Campbell, W. F., and Seeley, S. D.,** Temperature-stress-induced production of abscisic acid and dihydrophaseic acid in warm- and cool- season crops. *J. Am. Soc. Hortic. Sci.,* 106, 11, 1981.
15. **Chen, H.-H and Li, P. H.,** Potato cold acclimation, in *Plant Cold Hardiness and Freezing Stress: Mechanisms and Crop Implications,* Vol. 2, Li, P. H. and Sakai, A., Eds., Academic Press, New York, 1982, 5.
16. **Irving, R. M. and Lamphear, F. O.,** Regulation of cold hardiness in *Acer negundo, Plant Physiol.,* 43, 9, 1968.
17. **Gusta, L. V., Fowler, D. B., and Tyler, N. J.,** The effect of abscisic acid and cytokinins on the cold hardiness of winter wheat. *Can. J. Bot.,* 60, 301, 1982.
18. **Larson, A.,** unpublished information.
19. **Rikin, A. and Richmond, A. E.,** Amelioration of chilling injuries in cucumber seedlings by abscisic acid, *Physiol, Plant.,* 38, 95, 1976.
20. **Chen, T. H.-H. and Gusta, L. V.,** Abscisic acid-induced freezing resistance in cultured plant cells, *Plant Physiol.,* 73, 71, 1983.
21. **Orr, W., Singh, J., and Brown, D. C. W.,** Induction of freezing tolerance in alfalfa cell suspension cultures, *Plant Cell Rep.,* 4, 15, 1985.
22. **Kieth, C. N. and Mckersie, B. D.,** The effect of abscisic acid on the freezing tolerance of callus cultures of *Lotus corniculatus* L., *Plant Physiol.,* 80, 766, 1986.
23. **Reid, D. M., Pharis, R. P., and Roberts, D. W. A.,** Effects of four temperature regimens on the gibberellin content of winter wheat cv. Kharkov., *Physiol. Plant,* 30, 53, 1974.
24. **Proebsting, E. L., Jr. and Mills, H. H.** Gibberellin-induced hardiness responses in Elberta peach flower buds, *Proc. Am. Soc. Hortic, Sci.,* 85, 134, 1964.
25. **Edgerton, L. J.,** Some effects of gibberellin and growth retardents on bud development and cold hardiness of peach, *Proc. Am. Soc. Hortic. Sci.,* 88, 197, 1967.
26. **Andrews, J. E.,** Cold hardiness of sprouting wheat as affected by duration of hardening and hardening temperature, *Can. J. Plant Sci.,* 40, 94, 1960.
27. **Wilmar, J. C. and Doornbos, T.,** Stability of abscisic acid isomers to heat sterilization and light, in *Effects of Sterilization on Components in Nutrient Media,* van-Bragt, J., Mossel, D. A. A., Pierik, R. L. M., and Veldstra, H., Waggeningen, Netherlands, 1971, 139.
28. **Milborrow, B. V. and Vaughn, G.,** The long term metabolism of (±)-[^{14}C]-abscisic acid by apple seed, *J. Exp. Bot.,* 30, 983, 1979.
29. **Hedden, P., MacMillan, J., and Phinney, B. O.,** The metabolism of the gibberellins, *Ann. Rev. Plant Physiol.,* 29, 149, 1978.
30. **Rikin, A., Waldman, M., Richmond, A. E., and Dovrat, A.,** Hormonal regulation of morphogenesis and cold resistance. I. Modifications by abscisic acid and gibberellic acid in alfalfa (*Medicago sativa* L.) seedlings, *J. Exp. Bot.,* 26, 175, 1975.
31. **Vasil-Yev, I. M.,** *Wintering of Plants,* English translation, Levitt, J., Ed., Roger & Roger, Washington, D.C., 1961, 150.
33. **Sugawara, Y. and Sakai, A.,** Cold acclimation of callus culture of jerusalem artichoke, in *Plant Cold Hardiness and Freezing Stress: Mechanisms and Crop Implications,* Vol. 2, Li, P. H. and Sakai, A., Eds., Academic Press, New York, 1982, 541.

34. **Holubowitz, T., Pieniazek, J., and Khamis, M. A.**, Modification of frost resistance of fruit plants by applied growth regulators, in *Plant Cold Hardiness and Freezing Stress: Mechanisms and Crop Implications*, Vol. 2, Li, P. H. and Sakai, A., Eds., Academic Press, New York, 1982, 541.
35. **Yu, Y.-B. and Yang, S. F.**, Auxin-induced ethylene production and its inhibition by aminoethylomylglycine and cobalt ion, *Plant Physiol.*, 64, 1074, 1979.
36. **Erner, Y. and Jaffe, M. J.**, Thigmomorphogenesis: the involvement of auxin and abscisic acid in growth retardation due to mechanical perturbation. *Plant Cell Physiol.*, 23, 935, 1982.
37. **Titov, A. F., Drozdov, S. N., Kritenko, S. P., Talanova, V. V., Sherudilo, E. G.**, The effect of cytokinins on the cold and heat resistance of actively vegitating plants, *Fiziol. Biokhim. Kult. Rast.*, 18, 64, 1986.
38. **Kuraishi, S., Tezuka, T., Ushijima, T., and Tazaki, T.**, Effect of cytokinins on frost hardiness, *Plant Cell Physiology.*, 7, 705, 1966.
39. **Carter, J. V. and Brenner, M.**, Plant growth regulators and low temperature stress, in *Encyclopedia of Plant Physiology*, New series, Vol. 11, Pharis, A. P. and Reid, D. M., Eds., 1985, 418.
40. **Kacperska-Palacz, A., Blaziak, M., and Wcislinska, B.**, The effect of growth retardants CCC and B-9 on certain factors related to cold acclimation of plants, *Bot. Gaz. (Chicago)*, 130, 213, 1969.
41. **Gusta, L. V.**, unpublished information.
42. **Wilkinson, R. E.**, Mefluidide inhibition of *Sorghum* and gibberellin precursor biosynthesis, *J. Plant Growth Reg.*, 1, 85, 1982.
43. **Tseng. M. J. and Li, P. H.**, Mefluidide protection of severely chilled crop plants, *Plant Physiol.*, 75, 249, 1984.
44. **Zhang, C.-L., Li, P. H., and Brenner, M. L.**, Relationship between mefluidide treatment and abscisic acid metabolism in chilled corn leaves, *Plant Physiol.*, 81, 699, 1986.
45. **Rosa, J. T.**, Investigation on the hardening process in vegetable plants, *Mo. Agric. Exp. Stn. Res. Bull.*, 48, 1, 1921.
46. **Whiteside, A. G. O.**, Effect of soil drought on wheat plants, *Sci. Agric.* 21, 320, 1941.
47. **Chen, P. M., and Li, P. H.** Induction of frost hardiness in stem cortical tissues of *Cornus stolonifera* Michx. by water stress, *Plant Physiol.*, 59, 240, 1977.
48. **Boussiba, S., Rikin, A., and Richmond, A. E.**, The role of abscisic acid in cross-adaptation of tobacco plants, *Plant Physiol.*, 56, 337, 1975.
49. **Rikin, A. and Richmond, A. E.**, Factors affecting leakage from cucumber cotyledons during chilling stress, *Plant Sci. Lett.*, 14, 263, 1979.
50. **Mizrahi, Y., Blumenfeld, A., Bittner, S., and Richmond, A. F.**, Abscisic acid and cytokinen content of leaves in relation to salinity and relative humidity, *Plant Physiol.*, 48, 752, 1971.
51. **Christersson, L.**, Frost hardiness development in *Pinus silvestrus* L. seedlings at different levels of potassium and calcium fertilization, *Can. J. For. Res.*, 5, 738, 1975.
52. **Cook, T. W. and Duff, D. T.**, Effects of K fertilization on the freezing tolerance and carbohydrate content of *Festuca arundinacea* Schreb. maintained as turf *Agron. J.*, 68, 116, 1976.
53. **Beattie, J. H. and Flint, H. L.**, Effect of K level on frost hardiness of stems of *Forsythia intermedia* Zab. Lynwood, *J. Am. Soc. Hortic. Sci.* 98, 538, 1973.
54. **Mathias, E. L., Bennet, O. L., and Lundberg, P. E.**, Effect of rates of nitrogen on yield, nitrogen use and winter survival of midland bermudagrass *Cynodon dactylon* L. Pers. in Appalacia, *Agron. J.*, 65, 67, 1973.
55. **Pellet, N. E.**, Influence of nitrogen and phosphorus fertility on cold acclimation of roots and stems of two container-grown woody plant species, *J. Am. Soc. Hortic. Sci.*, 98, 82, 1973.
56. **Tyler, N. J., Gusta, L. V., and Fowler, D. B.**, The effect of water stress on the cold hardiness of winter wheat, *Can. J. Bot.*, 59, 1717, 1981.
57. **Ballarin-Denti, A. and Cocucci, M.**, Effects of abscisic acid, gibberellic acid, and fusicoccin on the transmembrane potential during the early phases of germination in radish *(Raphanus sativus* L.) seeds, *Planta*, 146, 19, 1979.
58. **Cocucci, S. and Cocucci, M.**, Effect of ABA, GA_3 and FC on the development of potassium uptake in germinating radish seeds, *Plant Sci. Let.*, 10, 85, 1977.
59. **Stevenick, van R. F. M.**, Abscisic acid stimulation of ion transport and alteration in $K+/Na+$ selectivity, *Z. Pflanzenphysiol.*, 67, 282, 1971.
60. **Lado, P., Rasi-Caldogno, F., and Colombo, R.**, Acidification of the medium associated with normal and fusicoccin induced seed germination, *Physiol. Plant.*, 34, 359, 1975.
61. **Malek, T. and Baker, D. A.**, Effect of fusicoccin on proton cotransport of sugars in the phloem loading of *Ricinus communis* L., *Plant Sci. Lett.*, 11 233, 1978.
62. **Markhart, A. H., III, Fiscus, E. L., Naylor, A. W., and Kramer, P. J.**, Effect of abscisic acid on root hydraulic conductivity, *Plant Physiol.*, 64, 611, 1979.
63. **Glinka, Z. and Reinhold, L.**, Abscisic acid raises the permeability of plant cells to water, *Plant Physiol.*, 48, 3, 1971.

64. **Glinka, Z.**, Effects of abscisic acid and of hydrostatic pressure gradient on water movement through excised sunflower roots, *Plant Physiol.*, 59, 933, 1977.
65. **Collins, J. C. and Morgan, M.**, The influence of temperature on the abscisic acid stimulated water flow from excised maize roots, *New Phytol.*, 84, 19, 1980.
66. **Tillberg, E., Holmvall, M., and Ericsson, T.**, Growth cycles in *Lemna gibba* cultures and their effects on growth rate and ultrastructure, *Physiol. Plant.*, 46, 5, 1979.
67. **Neskovic, M., Petrovic, J., Radojevic, L. J., and Vujicic, R.**, Stimulation of growth and nucleic acid biosynthesis at low concentrations of abscisic acid in tissue culture of *Spinacia oleracea, Physiol. Plant.*, 39, 148, 1977.
68. **Kochhar, T. S.**, Effect of abscisic acid and auxins on the growth of tobacco callus, *Z. Pflanzenphysiol.*, 97, 1, 1980.
69. **Overbeek, J. van. and Mason, M. I. R.**, Dormin and cytokinin: growth regulation of *Lemna, Acta Bot. Neerl.*, 17, 441, 1968.
70. **McLaren, J. S. and Smith, H.**, Effect of abscisic acid on photosynthetic products of *Lemna minor, Phytochemistry*, 16, 219, 1976.
71. **Aspinall, D., Paleg, L. G., and Addicott, F. T.**, Abscisin II and hormone regulated plant responses, *Aust. J. Biol. Sci.*, 20, 869, 1967.
72. **Blumenfeld, A. and Gazit, S.**, Interaction of kinetin and abscisic acid in the growth of soybean callus, *Plant Physiol.*, 45, 535, 1970.
73. **Aspinall, D.**, Role of abscisic acid and other hormones in adaptation of plants to water stress, in *Adaptation of Plants to Water and High Temperature Stress*, Turner, N. C. and Kramer, P. J., Eds., John Wiley & Sons, New York, 1980, 155.
74. **Stewart, C. R.**, The mechanism of abscisic acid induced proline accumulation in barley leaves, *Plant Physiol.*, 66, 230, 1980.
75. **McLaren, J. S., and Smith, H.**, The effect of abscisic acid on growth, photosynthetic rate and carbohydrate metabolism in *Lemna minor* L., *New Phytol.*, 76, 11, 1976.
76. **Heikkila, J. J., Papp, J. E. T., Schultz, G. A., and Bewley, J. D.**, Induction of heat shock protein messenger RNA in Maize mesocotyl by water stress, abscisic acid and wounding, *Plant Physiol.*, 76, 270, 1984.
77. **Gayler, K. R., and Glaziou, K. T.**, Plant enzyme synthesis: hormonal regulation of invertase and peroxidase synthesis in sugar cane, *Planta*, 84, 185, 1969.
78. **Glaziou, K. T. and Gayler, K. R.**, Storage of sugars in the stalks of sugar cane, *Bot. Rev.*, 38, 471, 1972.
79. **Mansfield, T. A. and Jones, R. J.**, Effects of abscisic acid on potassium uptake and starch content of stomatal guard cells, *Planta*, 101, 147, 1971.
80. **Newton, R. J.**, Abscisic acid effects on growth and metabolism in the roots of Lemna minor, *Physiol. Plant.*, 30, 108, 1974.
81. **Chin, C. C. and Weston, G. D.**, The relationship between invertase activity and growth in excised *Lycopersicon esculentum* roots, *Plant Sci. Lett.*, 4, 25, 1975.
82. **Watson, B. and Halperin, W.**, Reinvestigation of the effects of hormones and sugars on xylogenesis in cultured jerusalem artichoke *(Helianthus tuberosis*) tuber slices, with particular emphasis on the effects of different methods of media preparation and tissue analysis, *Z. Pflanzenphysiol.*, 101, 145, 1981.
83. **Soliday, C. L., Dean, B. B., and Kolattukudy, P. E.**, Suberization: inhibition by washing and stimulation by abscisic acid in potato disks and tissue culture, *Plant Physiol.*, 61, 170, 1978.
84. **Levitt, J.**, Chilling, freezing and high temperature stresses, in *Responses of Plants to Environmental Stresses*, 2nd Ed. Vol. 1, Academic Press, New York, 1980.
85. **Milborrow, B. V.**, The chemistry and physiology of abscisic acid. *Annu. Rev. Plant Physiol.*, 25, 258, 1974.
86. **Ackerson, R. C.**, Regulation of soybean embryogenesis by abscisic acid. *J. Exp. Bot.*, 35, 403, 1984.
87. **Takahashi, K.**, Abscisic acid as a stimulator for rice mesocotyl growth, *Nature. New Biol.*, 238, 92, 1972.
88. **Siminovitch, D., Gfeller, F., and Rheaume, B.**, The multiple character of the biochemical mechanism of freezing resistance of plant cells, cellular freezing and resistance in organisms, *The Institute of Low Temperature Science*, Asahina, E., Ed., Hokkaido University, Sapporo, Japan, 1967, 93.
89. **Kacperska-Palacz, A., Slugsbeka, E., Brectenwold, J., and Wcislinka, B.**, Physiological mechanisms of frost tolerance: possible role of protein in adaptation to cold, *Biol. Plant.*, 19(1), 10, 1977.
90. **Trunova, T. I. and Zvereva, G. H.**, Effect of protein synthesis inhibitors on frost hardiness of winter wheat. *Sov. Plant Physiol.*, 24, 311, 1977.
91. **Uemura, M. and Yoshida, S.**, Involvement of plasma membrane alterations in cold acclimation of winter rye seedlings *(Secale cereale* L.cv. Puma), *Plant Physiol.*, 75, 818, 19
92. **Yoshida, S and Uemura, M.**, Protein and isolated lipid compositions of isolated plasma membranes from orchardgrass *(Dactylis glomerata* L.) and changes during cold acclimation, *Plant Physiol.*, 75, 31, 1984.

93. **Guy, C. L., Nilmi, K. J., and Brambl, R.,** Altered gene expression during cold acclimation of spinach, *Proc. Natl. Acad. Sci. U.S.A.*, 82, 3673, 1985.
94. **Mozer, T. J.,** Control of protein synthesis in barley aleurone layers by the plant hormones gibberellic acid and abscisic acid, *Cell*, 20, 479, 1980.
95. **Triplett, B. A. and Quatrano, R. S.** Timing, localization and control of wheat germ agglutinin in developing wheat embryos, *Dev. Biol.*, 91, 491, 1982.
96. **Yoshida, S.,** Chemical and biophysical changes in the plasma membrane during cold acclimation of mulberry bark cells *(Morus bombycis* Koidz. cv. Goroji) *Plant Physiol.*, 76, 257, 1984.
97. **Uemura, M., and Yoshida, S.,** Involvement of plasma membrane alterations in cold acclimation of winter rye seedlings *(Secale cereale* L. cv. Puma.), *Plant Physiol.*, 75, 818, 1984.
98. **Ishikawa, M. and Yoshida, S.,** Seasonal changes in plasma membranes and mitochondria isolated from Jerusalem artichoke tubers. Possible relationship to cold hardiness, *Plant Cell Physiol.*, 26, 1331, 1985.
99. **Fowler, B. B., Gusta, L. V., and Tyler, N. J.,** Selection for winterhardiness in wheat. III. Screening methods, *Crop Sci.*, 21, 896, 1981.
100. **Dofing, S. M. and Schmidt, J. W.,** Inheritance of subcrown internode length in a winter barley cross, *Crop Sci.*, 24, 692, 1984.
101. **Dvorak, J. and Fowler, D. B.** Cold hardineess potential of triticale and tetraploid rye, *Crop Sci.*, 17, 477, 1978.

Chapter 2

METABOLIC CONSEQUENCES OF LOW TEMPERATURE STRESS IN CHILLING-INSENSITIVE PLANTS

Alina Kacperska

TABLE OF CONTENTS

I. Introduction 28

II. Frost Hardening Pattern in Winter Rape 28

III. Modification of Metabolism During Acclimation of Plants to Cold 29
- A. Metabolic Events Observed in Plants Grown in Cold 29
- B. Changes in Pyridine and Adenine Nucleotide Levels During Cold Acclimation and Their Possible Relations to Glucose Catabolism 33
- C. Two Phases of Plant Metabolic Response to Low Temperature 34

IV. Metabolic Events Associated with the Development of Freezing Tolerance 35

V. Summary 37

Acknowledgments 37

References 38

I. INTRODUCTION

Chilling-sensitive plants suffer from metabolic disfunction and structural injuries when exposed to chilling (>0°C) temperature; in contrast, the prolonged exposure of chilling-insensitive herbaceous plants to cold leads to two distinct effects: (1) improved performance of cells at low temperature and (2) increased resistance to freezing. These acclimation phenomena certainly are conditioned upon genetically fixed structural and physiological properties of cell constituents, termed adaptations.[1]

Improved performance of herbaceous plant cells at low temperature depends on two interrelated types of plant responses to the temperature change: an adjustment of metabolism and an adjustment of growth. The latter response has been thoroughly studied by Huner and his co-workers[2] who examined the morphological, anatomical, and biochemical consequences of leaf growth and development in cold. In our studies, we tried to differentiate between these two types of plant response in order to see what are the direct, metabolic responses of chilling insensitive plants to a sudden temperature decrease in the environment and in what way they can influence the long-termed (developmental) acclimation processes. Furthermore, we wanted to see what are the necessary requirements for the induction of freezing tolerance in the cold-acclimated tissues.

Most of our studies were performed on winter rape (*Brassica napus* L. var. *oleifera* L.) plants which were grown and hardened under controlled light and temperature conditions. In some cases, the experiments were performed on outdoor grown plants. After 5 weeks of growth at ambient temperature, plants were exposed to continuous low temperature (2°C) for at least 2 weeks. Analyses were performed on the young leaves which had begun to expand their blades before the cold-hardening took place. In some cases, other tissues (hypocotyls or roots) were also used. The cold-acclimated plants were subsequently exposed to subfreezing (<0°C) treatment to induce additional frost hardiness.[3]

II. FROST HARDENING PATTERN IN WINTER RAPE

Studies performed on the outdoor grown plants[3,4] showed that hardening of winter rape plants proceeds in three stages (Figure 1).

The first stage is induced by low but above 0°C temperature and results in increased frost resistance of only a few degrees below its initial level. From our most recent studies[5] it appeared that this stage can be observed in tissues capable for the extension growth, (e.g., young leaves) but it is not seen in tissues which had not yet entered or had already accomplished that stage of development before hardening occurred, (e.g., in apical buds or in hypocotyls, respectively). Differential thermal analysis (DTA) showed that freezing resistance of tissues being acclimated to cold relies manily on the supercooling ability of plant tissue water; beginning of ice formation as indicated by a single exotherm around −9 or −10°C resulted always in the death of more than 50% of cells, as checked with the conductivity method.[5] The studies also showed that a small increase in freezing resistance of the cold-grown winter rape leaves was mainly due to the decreased susceptibility of the tissue to the heterogenous ice nucleation induced by freezing of water on the leaf surface.[5]

The above observations confirm our previous hypothesis that the first stage of plant hardening results in the functional adaptation of metabolically active tissues to cold (>0°C), rather than in the increased tolerance of freezing.[3,6] Similar conclusions were recently made by Huner[2] for winter rye. Therefore, that step of frost hardening in herbaceous plant can be considered as the stage of plant acclimation to cold.

The second stage of frost hardening of winter rape plants depends on the temporary occurrence of subfreezing temperature, i.e., temperature lower than 0°C but higher than that of ice initiation in the tissue.[3,6] It results in a sudden increase in freezing tolerance in different

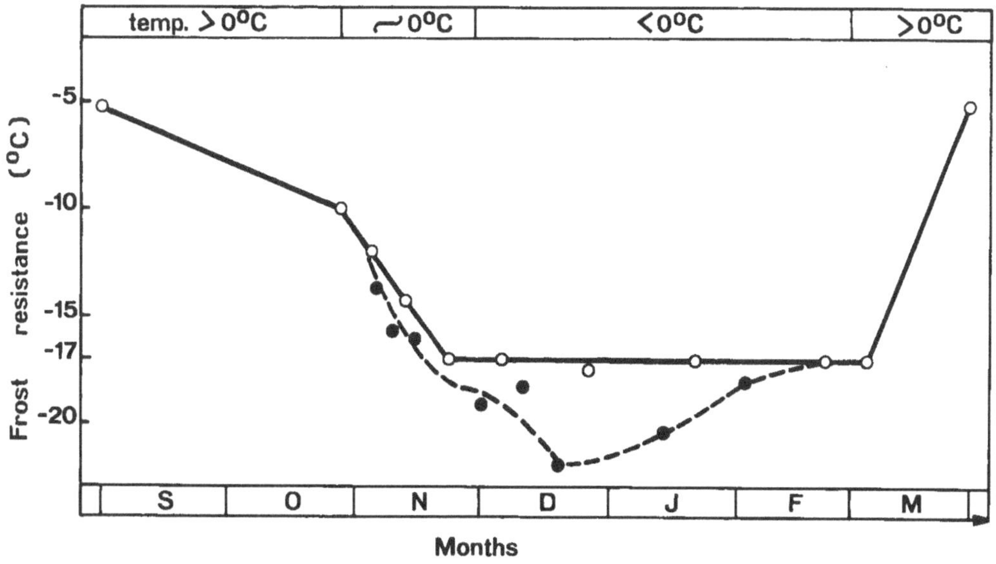

FIGURE 1. Three stages of frost acclimation in winter rape plants. Open and full dots indicate frost tolerance (T_{k50}) of rewatered and frost-dehydrated tissues, respectively. (From Kacperska-Palacz, A., *Plant Cold Hardiness and Freezing Stress*, Li, P. H. and Sakai, A., Eds., Academic Pres, New York, 1978. With permission.)

above-ground organs[5] and may result in a maximum frost tolerance in rewatered tissues[4] if the preceding light and temperature conditions were favorable for the plant acclimation to cold.[4] The prerequisite for subfreezing treatment for development of frost tolerance was also observed in other herbaceous plants.[7,8] Therefore, that step of plant hardening may be considered as the stage of a real frost tolerance development.

The third stage of frost hardening in winter rape plants is related to dehydration of cells by prolonged frosts and may overlap the second one if there are other water stress-provoking factors.[3]

It may be assumed that metabolic processes underlying each of the frost hardening stages in herbaceous plants are different and ought to be discussed separately.

III. MODIFICATION OF METABOLISM DURING ACCLIMATION OF PLANTS TO COLD

A. Metabolic Events Observed in Plants Grown in Cold

The expansion of the leaf blade is abruptly suspended upon a temperature drop to about 0°C and it takes several days in cold before the leaf growth can be resumed.[9] The suspension of cell enlargement by low temperature immediately results in such commonly observed events as accumulation of sugars, water soluble proteins, and increased dry matter content.[3] These events are frequently noted in tissues exposed to growth inhibiting agents other than low temperature, (e.g., growth inhibitors or retardants, specific light conditions, etc). Therefore, they seem to be a consequence of metabolic shift brought about by growth cessation, rather than by low temperature itself.[3] However, the accumulation of simple sugars is more profound in the cold-grown leaves than in the leaves pretreated with growth retardants;[10] this certainly is due to the cold-induced hydrolysis of polysaccharides.[11]

Among other modifications of metabolism observed in the cold-grown plants, inhibition of CO_2 assimilation and increased capacity of dark respiration are frequently noted. Our recent studies show that low temperature brings about two types of effects on plants with regard to photosynthesis and respiration: a direct response consisting of the rapid and readily

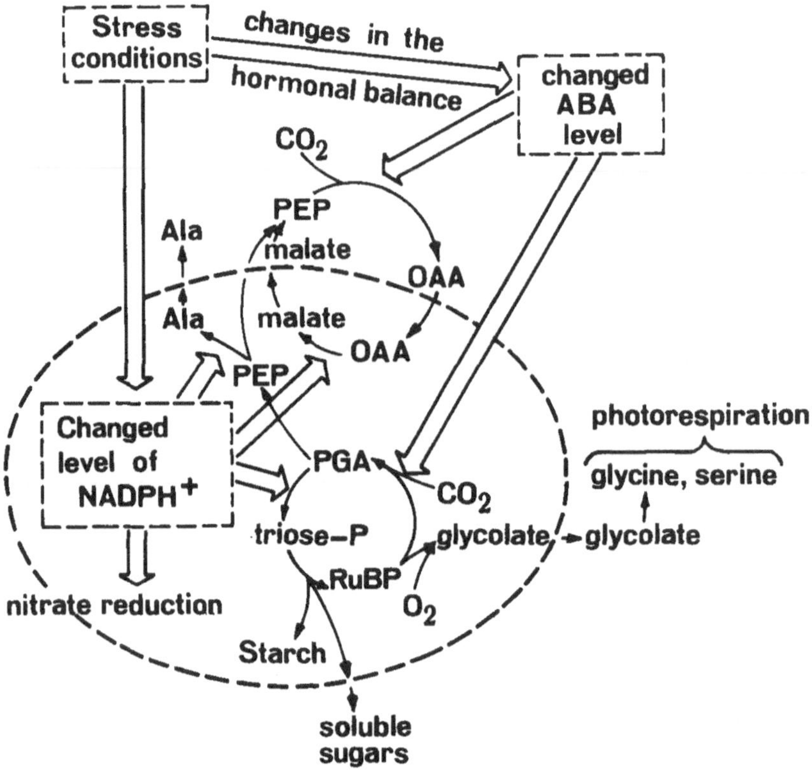

FIGURE 2. Proposed sites of stress action on photosynthetic carbon metabolism. The broken line indicates the chloroplast envelope. Abbreviations: ABA, abscisic acid; Ala, alanine; NADPH, nikotinamidedinucleotide phosphate; PEP, phosphoenolypyruvate; PGA, phosphoglyceric acid; OAA, oxalacetic acid; RuBP, ribulosebisphosphate.[19]

reversible reduction of both photosynthesis and respiration according to thermodynamic laws; and an indirect one which consists of long lasting modifications of photosynthesis and dark respiration rates and may be related to the cold-induced changes in the structural and biochemical properties of photosynthetic and respiratory systems.[12] These two types of modifications of photosynthesis and dark respiration can be regarded as functional and developmental adjustments, respectively.

Studies with ^{14}C incorporation into different photosynthates showed that, under cold conditions, synthesis of more reduced compounds such as alanine, malate, aspartate, glutamate, and proline dominated over synthesis of sugars.[13,14] After a short period (2 days) of plant exposure to cold, promoted synthesis of glycine and serine was also observed.[13] This effect disappeared in plants being acclimated to low temperatures for a longer period.[14] This again points out different cold-induced modifications of photosynthesis in plants responding or being acclimated to cold.

The relative predominance of ^{14}C incorporation to certain photosynthetic intermediates was also observed in plants exposed to stresses other than low temperature: heat,[15] water deficiency,[16] salinity,[17] and improper germination conditions.[18] Therefore, it can be assumed that basic mechanisms involved in the stress-induced modifications of photosynthetic carbon metabolism are common for different stresses with different herbaceous plants. It has been proposed that stress conditions promote these photosynthetic pathways which depend on a high availability of NADPH and respond to the increased ABA/GA ratio,[19] (Figure 2).

It is generally accepted that abscisic acid (ABA) level increases in tissues exposed to different stresses.[20] ABA, promoting the activity of PEP carboxylase and inhibiting that of RuBP carboxylase,[21] stimulates the carbon flow through C_4-dicarboxylic acids and restricts the activity of Calvin cycle. The increased ratio of PEPC activity to that of RuBPCase was actually observed in the cold-grown winter rape leaves.[22] Moreover, PEPC from wheat exhibited a low K_m (PEP) at low temperature[23] indicating a high affinity of the enzyme for this substrate in cold.

The other factor which affects the pathways of carbon assimilation in the stressed plants is the accessibility of NADPN, which results from the equilibrium between reactions producing and utilizing the reduced pyridine nucleotide. When light reactions are not restricted as much as dark reactions of photosynthesis, the production of NADPH may exceed the demand for the reaction of the Calvin cycle. Although there is no doubt that the rate of CO_2 assimilation decreases in cold, there is a controversy concerning the efficiency of photosynthetic electron transport in the cold-affected chloroplasts. In chloroplasts isolated from cold-hardened and over-wintering pine and spruce needles, whole chain, PSI, and PSII activities declined on per milligram chlorophyll basis.[24] However, Huner[25] showed that thylakoids from cold-hardened winter rye seedlings exhibited 40% higher light-saturated rates for whole chain electron transport than nonhardened thylakoids. The effect was primarily due to a 60% increase in capacity of PSI-associated electron transport in the hardened chloroplasts. According to Huner, the optimal isolation conditions for the hardened chloroplasts differ from those which are applied in studies of the nonhardened material and this might have been the main reason for a lower efficiency of the hardened chloroplasts when the suspension medium was the same for both types of chloroplasts (as it was in the studies on evergreen plants.[2] In in vivo experiments, no apparent photoinhibition of photosynthesis was observed in spinach leaves exposed for 40 hr to low temperature at either high or low light levels.[26] On the other hand, the photoinhibition observed in the fully expanded leaves, subjected to low temperature for a prolonged time, [27,28] seemed to be related to the light-promoted sensescence of these leaves because it was associated with a decrease in chlorophyll content. This could hardly be due to the cold-induced damage of the photosynthetic apparatus which is inflicted in a relatively short time at higher light fluency.

The relative surplus of NADPH in cold-hardened chloroplasts from chilling-resistant leaves can be utilized in reactions leading to the formation of compounds other than carbohydrates, e.g., alanine and malate (Figure 2). These metabolites are subsequently exported from chloroplasts and provide reducing equivalents and carbon skeletons for the reactions taking place outside chloroplasts, (e.g., malate being the substrate for the Krebs cycle instead of pyruvate). The increased NADPH availability also promotes nitrate reduction in the cold-affected leaves.[29]

Among metabolic events which are stimulated by low temperature in chilling insensitive tissues, accumulation of several amino acids[30,34] (alanine, glutamate and/or glutamine, aspartate and/or asparagine, serine and proline), accumulation of glutathione in its reduced form[35,38] and increased content of polyunsaturated fatty acids in lipids and phospholipids are frequently noted.[35,40] From the comparison of metabolic pathways and/or reactions promoted by cold with those which depend on the availability of the reduced pyridine nucleotides, (Figure 3) it follows that a high accessibility of these cofactors might be the "driving force" in the acclimation of chilling insensitive plants to cold, as previously proposed.[3,6]

The stimulation of some metabolic pathways by cold may have very profound and diverse biological consequences for the acclimation process. Apart from the possibility that the preferential synthesis of some amino acids may affect protein synthesis in cold, the important role of such compounds as proline, alanine, serine, and glycine in cellular osmoregulation[41,42] ought to be emphasized. Decreased water content in the cell is one of the first plant responses to chilling temperature[11] and osmoregulatory processes are thought to be associated with the maintenance of growth during water stress.[42] The accumulation of certain amino acids such

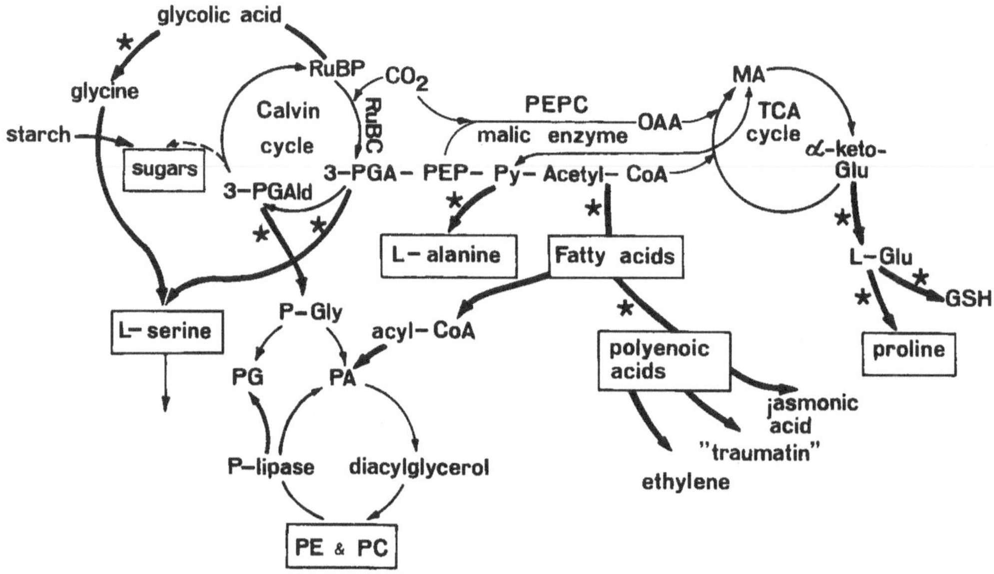

FIGURE 3. Comparison of some metabolic pathways dependent on the availability of reduced pyridine nucleotides (asteriks) and those promoted by cold (broad lines). Abbreviations: acetyl-CoA, acetyl-coenzyme A; acyl CoA, acyl-coenzyme A; L-Glu, L-glutamic acid; GSH, glutathione; α-keto Glu, α-ketoglutaric acid, MA, malic acid malate; OAA, oxalacetic acid; PA, phosphatidic acid, PC, phosphatidycholine; PE, phosphatidylethanolamine; PEP, phosphoenopyruvate; PEPC, phosphoenolpyruvate carboxylase; PG, phosphatidylglycerol; 3-PGA, 3-phosphoglyceric acid, 3-PGAld, glyceraldehyde-3-phosphate, P-Gly, glycerol phosphate; PK, pyruvate kinase; Py, pyruvate; RuBPC, ribulose-biphosphate carboxylase.[78]

as proline can also be of importance for the survival of freezing stress.[43] The accumulation of glutathione in its reduced form certainly is due to a high activity of NADPH dependent glutathione reductase in the cold-grown plants.[44] Glutathione participates in transhydrogenation reactions in the cell; it is involved in the formation and maintenance of protein thiol groups, provides reducing capacity for other reactions, (e.g., formation of deoxyribonucleotides), as well as protects proteins and cell membranes against peroxides and free radicals.[45] Although not associated with freezing tolerance[35,38] glutathione certainly is involved in the maintenance of cell functions at low temperature.

Interesting implications, arising from the hypothesis on the promotion of reducing equivalent–dependent metabolic pathways in cold, concern the fatty acid and lipid metabolism. The high level of unsaturation of lipids and phospholipids, reported frequently for the cold-affected, metabolically active tissue of herbaceous plants[46,47] but not for dormant tissues of woody perennials,[48] seems to be better understood now. Presumably, the high level of NADPH, as well as that of ATP promote biosynthesis of polyunsaturated fatty acids from acetylo-CoA in cold and the high level of reducing equivalents prevents their oxidation. The cold-promoted biosynthesis of linolenic acid from ^{14}C-acetate was actually noted during the first few days of cold treatment of winter wheat and rape roots,[49,50] whereas inhibition of linolenic acid and of phospholipid acyl chain catabolism was observed in the winter rape leaves exposed to cold for several days.[50] The high content of polyunsaturated fatty acids in phospholipids from the cold-grown herbaceous plants seems to be of importance for a better performance of membranes in cold, and it is not associated with a higher degree of freezing tolerance in cells.[47] It also may be an important factor in the induction of the second stage of hardening by subfreezing temperature, as discussed below.

The important role of "assimilatory power" in the metabolic processes occurring in cold was predicted long ago by Levitt.[51] He and his followers, however, sought evidence that

SH-groups of proteins are protected against the frost-induced denaturation by a high level of "reducing power" rather than look for other consequences of this phenomenon.

B. Changes in Pyridine and Adenine Nucleotide Levels During Cold Acclimation and Their Possible Relations to Glucose Catabolism

Experimental data on the changes in pyridine nucleotide levels and their redox state are rather scarce. Kuraishi et al.[52] showed that hardened pea plants maintained a higher NADPH content than the unhardened ones. Our recent experiments showed that not only NADPH but also NAD and NADP levels increased markedly in winter rape levels during the first few days of plant exposure to cold.[53] This finding indicates that not only reduction but also synthesis of pyridine nucleotides may be affected by cold. The leaves exposed to cold showed higher (by 70%) anabolic reduction charge and lower (by 44%) catabolic reduction charge. This observation confirms our previous supposition (Figure 3) that the cold-treated cells of chilling-insensitive plants possess a higher ability for reductive synthesis than the nonhardened plants. The question is whether the increase in NADPH level which followed that of NAD and NADP was only due to the cold-induced modifications of photosynthesis. Kuraishi et al.[52] observed that the elevated level of NADPH also was maintained in the cold-treated plants in darkness; this points out a possible involvement of pentose phosphate pathway in the formation of that compound. This possibility is in agreement with the results of Sagisaka[54] who showed a shift in glucose-6-phosphate metabolism from glycolysis to the pentose-phosphate cycle in poplar twigs in early autumn. Increased activity of glucose-6-phosphate dehydrogenase and important marked changes in the catalytic properties of pyruvate kinase also were observed in winter rape leaves exposed to cold.[55,56] Glucose-6-phosphate dehydrogenase was highly activated in *Chlorella elipsoidea* cells during the first 6 hr of cold treatment.[57] Therefore, the participation of pentose phosphate pathway in the formation of NADPH in the cold-treated tissues ought to be taken into consideration.

A high demand for ATP in the cold-acclimated, metabolically active herbaceous plants is actually fulfilled during the first few days of plant growth at low temperature; the level of ATP increased more than twice during 4 or 7 days of hardening of winter rape[58] and winter wheat[59] plants. No increase in ATP content was observed in the cold-treated winter rape roots[58] and in leaves of spring wheat leaves.[59] The frost hardening capacity of these tissues is very much limited.[58,59] None of the growth inhibiting factors such as ABA and water stress caused a rise in ATP level in the winter rape leaves. On the other hand, in chilling-sensitive cucumber leaves, the decrease in ATP level preceded the membrane permeability increase at low temperature.[61] From these observations it may be concluded that: (1) the ATP accumulation is specifically promoted by low temperature in tissues showing a capacity for frost hardening and (2) a high level of ATP may be needed not only for the maintenance of membrane integrity at low temperature but also it may be of importance for the energy-requiring processes occurring during the second stage of hardening (induced by subfreezing temperature). Also, its cryoprotective role for chloroplast membranes cannot be excluded.[62]

The promoted ATP formation by cold was observed not only in light but also in darkness.[58] Therefore, the participation of the ATP-generating systems other than photosynthesis in building up the ATP level in the cold-hardened tissues ought to be taken into consideration. Glycolysis involvement can hardly be anticipated because it is under control of a high adenylate energy charge and one of its key enzymes, pyruvate kinase, is strongly inhibited by ATP and alanine.[56] These negative modulators of pyruvate kinase are more effective in chilling-insensitive winter rape than in chilling-sensitive cucumber tissues.[56] Pyruvate kinase from nonacclimated winter rape leaves also showed a sharp increase in activation energy at temperature, whereas no such response was observed for pyruvate kinase from cucumber leaves.[56] It has been postulated that the catalytic properties of pyruvate kinase from winter rape leaves may represent an additional safeguard for glycolysis control in the cold-resistant tissues when other common regulatory systems fail.[56] Therefore, oxidative phosphorylation possibly takes part in the building up of high ATP level in cold-grown, chilling resistant plants. The phosphorylating efficiency of mitochondria and leaf segments from cereal plants grown in cold was shown to be high at low temperature,[63] even in the prefrozen material.[64] Winter rape plants exposed to cold showed a higher dark respiration capacity than the nonhardened plants; the most pronounced increase in respiratory activity was observed during the first few days of plant exposure to cold,[12] i.e., during the same period of time which was sufficient to markedly elevate the ATP level.[58] The increase in dark respiration rate during the first few days of cold treatment is due to the functioning of the KCN-resistant, alternative electron transport path in mitochondria of winter rape leaves.[65] That pathway is thought to operate in tissues having the cytochrome path inhibited or saturated with electrons when the supply of substrates for oxidation is high.[66] The marked increase in glucose content during the first few days of cold treatment was actually observed in winter rape leaves and it was experimentally proved that an addition of sucrose to the reaction medium increased the activity of the alternative electron path measured in the nonhardened leaf slices.[65] Therefore, it seems that the elevation of ATP content during the first few days of cold treatment is associated with the cold-increased supply of substrates for oxidation processes in concert with a high mitochondrial efficiency, the alternative path allowing for one additional phosphorylation when the cyrochrome path is fully saturated.[66,67]

The studies on pyridine and adenine nucleotide changes in the cold-grown leaves indicate that not only the light reactions of photosynthesis but also glucose metabolism play an important role in the acclimation process of herbaceous plants. That observation is in accordance with observations of other authors who repeatedly show a beneficial effect of simple sugars on hardening of seedlings. The effect certainly is due not only to their osmotic or cryoprotective action but also it is related to their role as substrates for the NADPH and ATP generating systems.

C. Two Phases of Plant Metabolic Response to Low Temperature

Data discussed above indicate that in young expanding leaf blades of chilling-insensitive plants dramatic changes in photosynthesis, dark respiration, and the energy state of a tissue take place during the first few days of cold treatment. Similar rapid and pronounced changes are observed at the enzymatic level; transient disturbances in activities of acid phosphatase, pyruvate kinase, and glucose-6-phosphate dehydrogenase were noted in winter rape leaves exposed for 4 days to low temperature.[55] The increase in glutathione reductase was also observed in other material.[44] Similar observations were made for the cold-promoted synthesis of linolenic acid in winter rape roots.[50] After several days of cold hardening, activities reached a new stable level,[12,50,55,68] which coincided with the reassumption of growth activity[9] and decreased (to the initial level) the anabolic reduction charge.[53] Therefore, we propose (Figure 4) that metabolic responses of plants to low temperature consist of two phases: (1) the reaction phase, when a sudden temperature change brings about disequilibrium in the cellular metabolism and (2) the restitution phase, when a new metabolic equilibrium is created which allows plants to function and grow under low temperature conditions. Similar biphasic physiological response was previously observed by Stocker[69] in plants subjected to water deficiency. Since in the second phase, not only recovery from the stress-induced disturbances but reorganization of metabolism and formation of new cell constituents, (i.e., synthesis of new proteins, lipids, nucleic acids, and other metabolites) take place,[11] the term "reconstruction phase" may better express the physiological meaning of the phase.

The ability of a plant to pass from the reaction to the reconstruction phase may constitute the basic difference between chilling-sensitive and chilling-insensitive plants and might be the prerequisite for a successful acclimation of plants to cold. The ability certainly depends

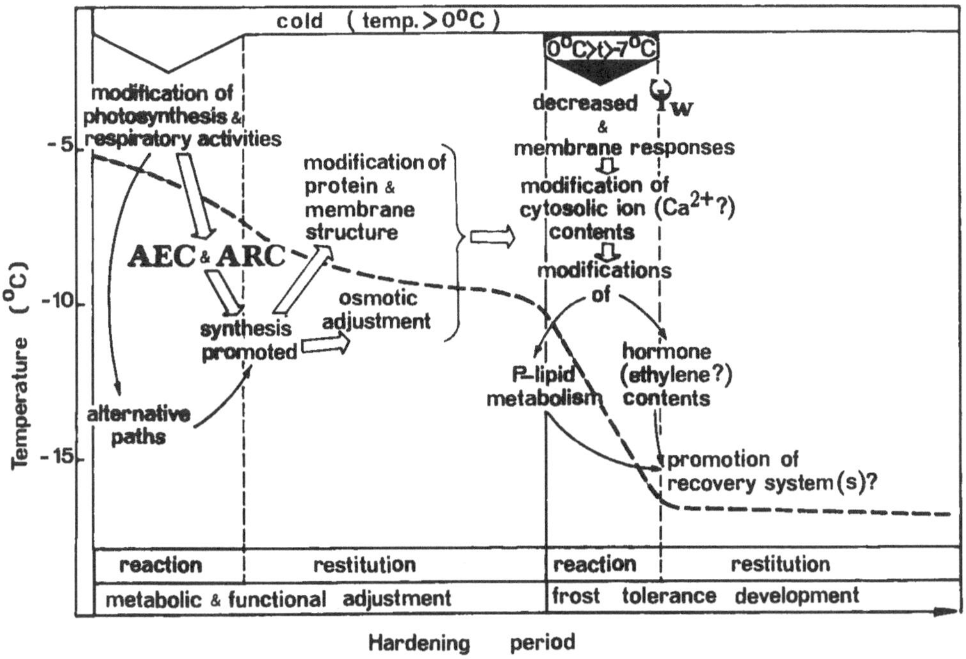

FIGURE 4. Metabolic responses of chilling-insensitive herbaceous plants to cold (temp.>0°C) or subfreezing (0>temp.> − 7°C) in the course of frost hardening. Broken line indicates frost resistnce (T_{k50}). Abbreviations: AEC, adenylate energy charge; ARC, anabolic reduction charge; ψ_w (Psi_w) water potential.

on plant constitutional adaptations, but also the existence of mechanisms responsible for the functional adjustment of cells is of great importance. As shown above, the energy metabolism of cells is strongly affected by cold, resulting in the elevation of ATP and reducing equivalent levels. The need for transduction or dissipation of energy in different cell compartments may, therefore, be anticipated. It is interesting that during the reaction phase, metabolic pathways such as glycolic acid (photorespiration)[14] and the cyanide-resistant respiratory path[65] showed the highest activity in winter rape leaves. Among its other functions, glycolic acid pathway may operate as the terminal oxidase system which efficiently uses the surplus of NADPH in chloroplasts[70] and also enables the formation of sucrose on the alternative pathway when sugar synthesis via the Calvin cycle is restricted.[70] It is also proposed that photorespiration under unfavorable environmental conditions may protect the photochemical centers not only by recycling of CO_2 but also directly, by maintaining the photosynthetic electron transport.[71] The operation of alternative electron path in mitochondria of chilling insensitive plants, under conditions of a high sugar supply, may enable "an energy or overcharge overflow"[64,67] and the functioning of TCA cycle supplies the cell with carbon skeletons for different synthesis. It is of interest that the cyanide-resistant path is proposed to play a role in removing excess of reducing power from a cell in cooperation with the malic enzyme (taking part in the oxidation of malate).[63]

It seems that the presence of both glycolic acid pathway and alternative electron pathway in chilling-insensitive tissues substantially contributes to the capacity of a plant to respond to a rapidly changing environment.

IV. METABOLIC EVENTS ASSOCIATED WITH THE DEVELOPMENT OF FREEZING TOLERANCE

In contrast with the relatively abundant knowledge on modifications of metabolisms during the first stage of plant acclimation to low temperature, there is much less information of the mechanisms involved in the acquisition of freezing tolerance by a cell.

There is no doubt that the plasma membrane is the primary site of freezing injury in higher plants.[74] The initial step in cold acclimation process is to decrease the sensitivity of that membrane to mechanical stress occurring during osmotic contraction and expansion caused by freezing/thawing.[74] Some of the metabolic events observed during the first stage acclimation, (e.g., accumulation of cryoprotectants and osmoregulatory agents of modified membrane composition and structure) certainly contribute to a higher resistance of plasma membrane and, possibly, other membranes to the frost-induced dehydration effects. However, as mentioned above, temperature lower than 0°C but higher than that of ice formation in the tissue is required to induce the development of freezing tolerance in chilling-insensitive biannual plants[3] and woody perennials.[75] The effect of subfreezing temperature is associated with an initial membrane permeability increase, concomitant with the transient decrease of phosphatidylcholine level and the elevation of phosphatidylglycerol content in the leaf tissue.[76] Initial alterations of osmotic properties of membranes were shown to promote the phosphatidyl transferase activity in the cold-acclimated winter rape leaves.[76,77] It has been proposed that the frost-induced increase in transferase activity of phospholipase D (in the presence of such alcohols as glycerol and ethanolamine) contributes to synthesis of phosphatidylcholine or phosphatidylethanolamine in cells and to remodelling of membrane structure.[77] Therefore, the hardening effect of slight frosts seems to consist of the activation of a recovery system in cells.[77] With this hypothesis are other consistent findings indicating that the initial modifications of membrane properties activate lipoxygenase (EC 1.13.11.12) which catalyzes the oxygenation of polyunsaturated fatty acids, forming fatty acid peroxides.[78] Lipoperoxides can be further metabolized to such biologically active substances as 12-oxo-*trans*-dodecenoic acid,[79] hypothetic "wound hormone" of Harberlandt or jasmonic acid,[80] a novel growth regulating substance. Formation of lipoperoxides was recently shown to be coupled with biosynthesis of ethylene, lipoxygenase being involved in the formation of ethylene from l-aminocyclopropane-l-carboxylic acid (ACC).[78] The exact role of these growth regulators in the development of frost tolerance is at present far from being understood.

Recently, it was shown that leaf discs from the cold-acclimated winter rape plants exhibited a marked decreased in water potential after pretreatment with subfreezing temperature. This was associated with the increased capacity for the leaf expansion growth.[81]

In view of the above facts, it seems that the altered osmotic behavior of cells pretreated with slight, transient frosts, cannot be considered as "sublethal manifestation of injury"[74] but is a prerequisite for development of freezing tolerance. It may be regarded rather as the manifestation of the other "reaction phase" which leads to the new "restitution or reconstruction phase" during plant hardening (Figure 4).

The nature of membrane modifications, caused by subfreezing temperature, is not known. In winter rape leaves, membrane alterations are not due to the lipid phase transition or separation because such a change in phospholipid properties occurs in winter rape at 0 or -2°C, depending on the state of acclimation of leaves to cold.[82] Neither is it caused by a sudden water potential disequilibrium associated with the formation of ice, because it occurs at a temperature higher than that of ice initiation.[5] It has to be pointed out that most of the responses described above are under the control of calcium ions. It is tempting to propose that a sudden alteration of calcium balance in the cytosol may modify the activity of calcium-dependent enzymes either directly, (e.g., phospholipase D or lipoxygenase activities) or indirectly, through the calmodulin-Ca^{2+} dependent system.[83,84] The verification of that hypothesis, as well as a search for possible reasons of changes in calcium balance in subfrozen cells, may open a new field in the studies on mechanisms of freezing tolerance in plants.

V. SUMMARY

Having recognized that acclimation of chilling-insensitive herbaceous plants to low temperature proceeds in three stages, it ought to be pointed out that the first stage relies mainly on the metabolic and functional adjustments of cells to cold, and it is not associated with the increased freezing tolerance. During that stage, a rapid response of the growing tissue to the decreased environmental temperature is observed and might be called ''the reaction phase'' (Figure 2). This phase consists mainly of a rapid increase of adenylate and anabolic reduction charges in cells (AEC and ARC, respectively), transient disturbances in several enzymes involved in glucose catabolism, and it results in the promotion of these metabolic pathways which depend on a high availability of ATP and reducing equivalents (Figure 3).

During the prolonged exposure of plants to cold ()0°C), a new metabolic equilibrium is created which allows plants to function and grow under low temperature conditions. This phase of plant adjustment to cold may be called ''the restitution phase''(Figure 4). Preferential synthesis of some photosynthates in cold (alanine, malate, glutamate, asparate, proline) may have important meaning for cellular osmoregulation and the maintenance of growth under cold conditions. The accumulation of glutathione in its reduced form (GSH) is of importance for the maintenance of cell functions at low temperature, although it is not associated with freezing tolerance. The same concerns the cold-induced changes in fatty acid and lipid metabolisn which lead to a high level of unsaturation of phospholipids in the cold-grown herbaceous plants. A high unsaturation of membrane lipids improve membrane performance at low temperature but is not necessarily related to a higher freezing tolerance of a tissue.

It seems that the ability of the plant to pass from the reaction to the restitution phase may be the prerequisite for a successful acclimation of plants to cold. Since the energy metabolism of cells is strongly affected by cold, the need for transduction or dissipation of energy in different cell compartments may, therefore, be anticipated. It is proposed that the glycolic acid pathway and cyanide-resistant electron path which operate during trhe reaction phase present the ''safety valves'' which protect the cold-affected cells against secondary effects of disturbances in the activities of energy generating systems.

The acquisition of freezing tolerance by the cold-acclimated cells depends on mechanisms induced by subfreezing temperature (temperature lower than 0°C but higher than that of ice initiation in a tissue). The nature of these mechanisms is poorly understood. They seem to be induced by small alterations in membrane semipermeable properties and to rely on the profound changes in phospholipid and fatty acid catabolism. These changes seem to lead to the activation of the recovery system in cells with phospholipase D and lipoxygenase being involved in the cell remodeling or in the formation of biologically active substances, respectively. It has been shown that the subfrozen cells exhibit a sudden decrease in water potential which is associated with the increased capacity for growth. The very nature of this phenomenon calls for further studies.

ACKNOWLEDGMENTS

The author is grateful for financial support from the Maria Sklodowska-Curie Fund (established by contribution of the U.S. and Polish Governments) and from Polish Academy of Sciences and Ministry of Higher Education.

REFERENCES

1. **Berry, J. A. and Björkman, O.,** Photosynthetic response and adaptation to temperature in higher plants, *Annu. Rev. Plant Physiol.*, 31, 491, 1980.
2. **Huner, N. P. A.,** Low temperature acclimation: leaf ontogeny and biochemistry, *Acta Physiol. Plant*, 8, 53, 1986.
3. **Kacperska-Palacz, A.,** Mechanism of cold acclimation in herbaceous plants, in *Plant Cold Hardiness and Freezing Stress*, Li, P. H. and Sakai, A., Eds., Academic Press, New York, 1978, 139.
4. **Sikorska, E. and Kacperska-Palacz, A.,** Phospholipid involvement in frost tolerance, *Physiol. Plant*, 47, 144, 1979.
5. **Kulesza, L., Pukacki, P., and Kacperska, A.,** Ice formation and frost killing temperatures as related to cold acclimation of winter rape plants, *Acta Physiol. Plant*, 3, 185, 1986.
6. **Kacperska, A.,** Biochemical and physiological aspects of frost hardening in herbaceous plants, in *Plant Production in the North*, Kaurin, Å., Junttila, O., and Nilsen, J., Eds., Norvegian University Press, Oslo, 1985, 99.
7. **Cox, W. and Levitt, J.,** Interrelations between environmental factors and freezing resistance of cabbage leaves, *Plant Physiol.*, 57, 553, 1976.
8. **Andrews, C. J., Pomeroy, M. K., and de la Roche, I. A.,** The influence of light and diurnal freezing temperature on cold hardiness of winter wheat seedlings, *Can. J. Bot.*, 52, 2539, 1974.
9. **Maciejewska, U.,** Personal communication, 1986.
10. **Kacperska-Palacz, A., and Wciślińska, B.,** The effect of CCC on the nitrogen compound contents in rape plants and their frost hardiness. Relation to the conditions of day-length and temperature, *Biol. Plant*, 14, 39, 1972.
11. **Levitt, J.,** *Responses of Plants to Environmental Stresses*, Vol. 1, 2nd ed., Academic Press, New York, 1980, chap. 8.
12. **Maciejewska, U., Tomczyk, J., and Kacperska, A.,** Effects of cold on CO_2 exchange in winter rape leaves, *Physiol. Plant*, 62, 315, 1984.
13. **Maciejewska, U., Maleszewski, S., and Kacperska-Palacz, A.,** Effect of cold hardening on the carbon photosynthetic metabolism in rape leaves, *Bull. Acad. Pol. Sci.* 22, 513, 1974.
14. **Sosińska, A., Maleszewski, S., and Kacperska-Palacz, A.,** Carbon photosynthetic metabolism in leaves of cold-treated rape plants, *Z. Pflanzenphysiol.*, 83, 285, 1977.
15. **Chljustova, T. M. and Tarchevsky, I. A.,** Effets of high temperature on ^{14}C incorporation to photosynthetic products in leaves and isolated chloroplasts of spinach, *Sov. Plant Physiol.*, 19, 251, 1972.
16. **Lawlor, D. W.,** Assimilation of carbon into photosynthetic intermediates of water-stressed wheat, *Photosynthetica*, 10, 431, 1976.
17. **Passera, C. and Albuzio, A.,** Effect of salinity on photosynthesis and photorespiration of two wheat species *Triticum durum* cv. PEPE 2122 and *Triticum aestivum* cv. Marzotto, *Can. J. Bot.*, 56, 121, 1978.
18. **Maciejewska, U. and Maleszewski, S.,** Embryonal dormancy and photosynthetic carbon metabolism in apple seedlings, *Z. Pflanzenphysiol.*, 79, 300, 1976.
19. **Kacperska, A., Lewak, S., Maciejewska, U., and Maleszewski, S.,** Photosynthesis control by environmental factors, *Pol. Ecol. Bibliogr.* 7, 377, 1981.
20. **Wright, S. T. C.,** Phytohormones and stress phenomena, in *Phytohormones and Related Compounds — A Comprehensive Treatise*, Letham, D. S., Goodwin, P. B., and Higgins, T. J. V., Eds., Elsevier, Amsterdam, 1978, 495.
21. **Sankhla, N. and Huber, W.,** Enzyme activities in *Pennistetum* seedlings germinated in the presence of abscisic and gibberelic acids, *Phytochemistry*, 13, 543.
22. **Sosińska, A. and Kacperska-Palacz, A.,** Ribulosediphosphate and phosphoenolpyruvate carboxylase activities in winter rape leaves as related to cold acclimation, *Z. Pflanzenphysiol.*, 92, 455, 1979.
23. **Graham, D., Huckley, D. G., and Patterson, R. D.,** Temperature effects on phosphoenolpyruvate carboxylase from chilling-sensitive and chilling-resistant plants, in *Low Temperature Stress in Crop Plants*, Lyons, J. M., Graham, D., and Raison, J. K., Eds., Academic Pres, New York, 1979, 453.
24. **Öquist, G.,** Effects of Low temperature on photosynthesis, *Plant Cell Environ.*, 6, 281, 1983.
25. **Huner, N. P. A.,** Acclimation of winter rye to cold hardening results in an increased capacity for photosynthetic electron transport, *Can. J. Bot.*, 63, 506, 1985.
26. **Garber, M. P.,** Effect of light and chilling temperatures on chilling-sensitive and chilling-resistant plants, *Plant Physiol.*, 59, 981, 1977.
27. **Ögren, E., Öquist, G., and Höllgren, J. E.,** Photoinhibition of photosynthesis in Lemma gibba as induced by the interaction between light and temperature. I. Photosynthetis in vivo, *Physiol. Plant*, 62, 181, 1984.
28. **Steffen, K. L. and Palta, J. W.,** Effect of light on photosynthetic capacity during cold acclimation in a cold-sensitive and a cold-tolerant potato species, *Physiol. Plant*, 66, 353, 1986.

29. **Weidner, M. and Salisbury, F. B.**, The temperature characteristics of ribulose-1,5,-diphosphate carboxylase, nitrate reductase, and pyruvate kinase from seedlings of two spring wheat varieties, *Z. Pflanzenphysiol.*, 71, 398, 1974.
30. **Draper, S. R.**, Amino acid changes associated with low temperature treatment of Lolium perenne, *Phytochemistry*, 11, 639, 1972.
31. **Chu, T. M., Jusaitis, M., Aspinall, D., and Paleg, L. G.**, Accumulation of free proline at low temperatures, *Physiol. Plant*, 43, 254, 1978.
32. **Özturk, M. and Szaniawski, R. K.**, Root temperature stres and proline content in leaves and roots of two ecologically different plant species, *Z. Pflanzenphysiol.*, 102, 375, 1981.
33. **Patterson, B. D., Pearson, J. A., Payne, L. A., and Ferguson, J. B.**, Metabolic aspects of chilling resistance: alanine accumulation and glutamate depletion in relation to chilling sensitivity in passion-fruit species, *Aust. J. Plant Physiol.*, 8, 395, 1981.
34. **Sagisaka, S. and Araki, T.**, Amino acid pools in perennial plants at the wintering stage and at the beginning of growth, *Plant Cell Physiol.*, 24, 479, 1983.
35. **Esterbauer, H. and Grill, D.**, Seasonal variation of glutathione and glutathione reductase in needles of *Picea abies*, *Plant Physiol.*, 61, 119, 1978.
36. **de Kok, L. J., de Kan, P. J. L., Tańczos, O. G., and Kuiper, P. J. C.**, Sulphate induced accumulation of glutathione and frost-tolerance of spinach leaf tissue, *Physiol. Plant*, 53, 435, 1981.
37. **de Kok, L. J. and Oosterhuis, F. A.**, Effects of frost-hardening and salinity on glutathione and sulfhydryl levels and on glutathione reductase activity in spinach leaes, *Physiol. Plant*, 58, 47, 1983.
38. **Guy, C. L., Carter, J. V., Yelenosky, G., and Suy, C. T.**, Changes in glutathione content during cold acclimation in *Cornus sericea* and *Citrus sinensis*, *Cryobiology*, 21, 443, 1984.
39. **Kuiper, P. J. C.**, Lipid in alfalfa leaves in relation to cold hardiness, *Plant Physiol.*, 45, 684, 1970.
40. **Grenier, G. and Willemot, C.**, Lipid changes in roots of frost hardy and less hardy alfalfa varieties under hardening conditions. *Cryobiology*, 11, 324, 1974.
41. **Hochacka, P. W. and Somero, G. N.**, *Strategies of Biochemical Adaptation*, W. B. Saunders, Philadelphia, 1973, chap. 4.
42. **Morgan, J. N.**, Osmoregulation and water stress in higher plants, *Annu. Rev. Plant Physiol.*, 35, 299, 1984.
43. **Withers, L. A. and King, P. J.**, Proline: a novel cryoprotectant for the freeze preservation of cultured cells of *Zea mays* L., *Plant Physiol.*, 64, 675, 1979.
44. **Guy, C. L. and Carter, J. V.**, Characterization of partially purified glutathione reductase from cold-hardened and non-hardened spinach leaf tissue, *Cryobiology*, 21, 454, 1984.
45. **Meister, A.**, Metabolism and function of glutathione, *Trend Biochem. Sci.*, 6, 231, 1981.
46. **Willemot, C., Hope, C. J., Hope, R. J., Williams, R. J., and Michand, R.**, Changes in fatty acid composition of winter wheat during frost hardening, *Cryobiology*, 14, 87, 1977.
47. **Smoleńska, G. and Kuiper, P. J. C.**, Effect of low temperature upon lipid and fatty acid composition of roots and leaves of winter rape plants, *Cryobiology*, 41, 29. 1977.
48. **Singh, J., de la Roche, I. A., and Siminovitch, D.**, Membrane augmentation in freezing tolerance of plant cells, *Nature (London)*, 257, 620, 1975.
49. **Grenier, G., Hope, H. J., Willemot, C., and Therrien, H. P.**, Sodium 1,2-^{14}C acetate incorporation in roots of frost hardy and less hardy alfalfa varieties under hardening conditions, *Plant Physiol.*, 55, 906, 1975.
50. **Smoleńska-Sym, G., and Kacperska, A.**, Phospholipid and fatty acid turnover in winter rape plants subjected to cold treatment, in *Proc. Int. Rapeseed Conf.*, Vol. 1, G. C. I. R. C., Paris, 83, 1983.
51. **Levitt, J.**, The mechanism of hardening on the basis of the SH$\rightleftharpoons$SS hypothesis of freezing injury, in *Cellular Injury and Resistance in Freezing Organisms*, Asahina, E., Ed., Inst. Low. Temp. Sci., Hokkaido Univ., Sapporo, Japan, 1967, 41.
52. **Kuraishi, S., Arai, N., Ushijima, T., and Tazaki, T.**, Oxidized and reduced nicotinamide adenine dionucleotide phosphate levels of plants hardened and unhardened against chilling injury, *Plant Physiol.*, 43, 238, 1968.
53. **Maciejewska, U. and Kacperska, A.**, Changes in the levels of oxidized and reduced pyridine nucleotides during cold acclimation of winter rape plants, *Physiol. Plant*, 69, 687, 1987.
54. **Sagisaka, S.**, Transition of metabolism in living poplar bark from growing to wintering stages and vice versa, *Plant Physiol.*, 54, 544, 1974.
55. **Sobczyk, E. A. and Kacperska-Palacz, A.**, Changes in some enzyme activities during cold acclimation of winter rape plants, *Acta Physiol. Plant* , 2, 123, 1980.
56. **Sobczyk, E. A., Rybka, Z., Kacperska, A.**, Modification of pyruvate kinase activity in cold-sensitive and cold-resistant leaf tissues, *Z. Pflanzenphysiol.*, 114, 285, 1984.
57. **Sadakane, H. and Hatano, S.**, Isoenzymes of glucose 6-phosphate dehydrogenase in relation to frost hardiness of *Chlorella elipsoidea*, in *Plant Cold Hardiness and Freezing Stress*, Vol. 2, Li, P. H. and Sakai, A., Eds., Academic Press, New York, 1982, 157.

58. **Sobczyk, E. A. and Kacperska-Palacz, A.,** Adenine nucleotides changes during cold acclimation of winter rape plants, *Plant Physiol.*, 62, 875, 1978.
59. **Perras, M. and Sarhan, F.,** Energy state of spring and winter wheat during cold hardening. Soluble sugars and adenine nucleotides. *Physiol. Plant*, 60, 129, 1984.
60. **Sobczyk, E. A.,** Involvement of Adenine Nucleotides and Some Enzyme Activities in Plant Responses to Low Temperature, Ph.D. thesis, Univ. of Warsaw, Poland, 1982
61. **Sobczyk, E. A., Marszalek, A., and Kacperska, A.,** ATP involvement in plant tissue responses to low temperature, *Physiol. Plant*, 63, 399, 1985.
62. **Santarius, K.,** Effective cryoprotection of thylakoid membranes by ATP, *Planta*, 161, 555, 1984.
63. **Pomeroy, M. K. and Andrews, C. J.,** Effect of temperature on respiration of mitochondria and shoot segments from cold-hardened and non-hardened wheat and rye seedlings, *Plant Pysiol.*, 56, 703, 1975.
64. **Singh, J., de la Roche, I. A., and Siminovitch, D.,** Relative insensitivity of mitochondria in hardened and non-hardened rye coleoptile cells to freezing in situ, *Plant Physiol.*, 60, 713, 1977.
65. **Rychter, A. and Ciesla, E., and Kacperska, A.,** Participation of the cyanide-resistant pathway in respiration of winter rape leaves as affected by plant cold acclimation, *Physiol. Plant*, 73, 299, 1988.
66. **Lance, C., Chauveau, M., and Dizengremel, P.,** The cyanide-resistant pathway of plant mitochondria, in *Encyclopedia of Plant Physiology* New Series, Vol. 18, Douce, R. and Day, D. A., Eds., Springer-Verlag, Berlin, 1985, 202.
67. **De Visser, R., Brouwer, K. S., and Posthumus, F.,** Alternative path mediated ATP synthesis in roots of *Pisum sativum* upon nitrogen supply, *Plant Physiol.*, 80, 295, 1986.
68. **Sobczyk, E. A., Shcherbakova, A., and Kacperska, A.,** Effect of cold acclimation on the stability of some enzymes in winter rape plants, *Z. Pflanzenphysiol.*, 100, 113, 1980.
69. **Stocker, O.,** Physiological and morphological changes in plants due to water deficiency, *Arid Zone Res.*, 15, 63, 1960.
70. **Ogren, W.,** Photorespiration: pathways, regulation and modification, *Annu. Rev. Plant Physiol.* 35, 415, 1984.
71. **Sowiński, P. and Zagdańska, B.,** Photorespiration as a ''safety valve'' for the light phase of photosynthesis under unfavourable environmental conditions, *Bull. Pol. Acad. Sci.*, 32, 43, 1984.
72. **Lambers, H.,** Cyanide-resistant respiration: a non-phosphorylating electron transport pathway acting as an energy overflow, *Physiol. Plant*, 55, 478, 1982.
73. **Rustin, P., Moreau, F., and Lance, C.,** Malate oxidation in plant mitochondria via malic enzyme and cyanide-insensitive electron transport pathway, *Plant Physiol.*, 66, 457, 1980.
74. **Steponkus, P.,** Role of the plasma membrane in freezing injury and cold acclimation, *Annu. Rev. Plant Physiol.*, 35, 543, 1984.
75. **Weiser, C. J.,** Cold resistance and injury in woody plants, *Science*, 169, 1269, 1970.
76. **Sikorska, E., and Kacperska-Palacz, A.,** Frost-induced phospholipid changes in cold-acclimated and non-acclimated rape leaves, *Physiol. Plant*, 48, 201, 1980.
77. **Sikorska, E. and Kacperska, A.,** Freezing-induced membrane alterations: injury or adaptation?, in *Plant Cold Hardiness and Freezing Stress*, Vol. 2, Li, P. H. and Sakai, A., Eds., Academic Press, New York, 1982, 261.
78. **Kacperska, A., and Kubacka-Zebalska, M.,** Is lipoxygenase involved in the formation of ethylene from ACC?, *Physiol. Plant.*, 63, 399, 1985.
79. **Zimmerman, D. C. and Coudron, C. A.** , Identification of traumatin, a wound hormone, as 12-oxo-trans-10-dodecenoic acid, *Plant Physiol.*, 63, 536, 1979.
80. **Vick, B. A. and Zimmerman, D. C.,** The biosynthesis of jasmonic acid: a physiological role for plant lipoxygenase, *Biochem. Biophys. Res. Commun.*, 111, 470, 1983.
81. **Kacperska, A. and Kulesza, L.,** Frost resistance of winter rape leaves as related to the changes in water potential and growth capability, *Physiol. Plant*, 71, 483, 1988.
82. **Sikorska, E., Ondrias, K., and Farkas, T.,** Physical properties of membrane of cold-hardened and non-hardened winter rape plants, *Acta Biol. Acad. Sci. Hungary*, 32, 267, 1981.
83. **Dieter, P.,** Calmodulin and calmodulin-mediated processes in plants, *Plant Cell Environ.*, 7, 371, 1984.
84. **Minorsky, P. V.,** An heuristic hypothesis of chilling injury in plants: a role for calcium as the primary physiological transducer of injury, *Plant Cell Environ.*, 8, 75, 1985.

Chapter 3

ALTERATIONS OF PLASMA MEMBRANES RELATED TO COLD ACCLIMATION OF PLANTS

Shizuo Yoshida and Matsuo Uemura

TABLE OF CONTENTS

I. Introduction 42

II. Lipid Changes 42

III. Protein Changes 44

IV. Alterations of Physical Properties 47

V. Summary 49

References 50

I. INTRODUCTION

There is mounting evidence to suggest that plasma membranes play a central role in freezing injury of plant cells.[1,2] Therefore, it is quite natural to assume that plasma membranes undergo chemical alterations during cold acclimation so as to adjust their properties against freezing stress. However, no direct evidence has been presented until now as to what chemical changes are produced on the plasma membranes and how the changes are responsible for their protection against freezing. A major problem encountered by investigators attempting to analyze plasma membrane compositions, however, is the difficulty of plasma membrane isolation in a high purity and in large quantities needed for the detailed studies. Recently, we have established a method to purify plasma membranes from most plants using an aqueous two-polymer phase system containing NaCl.[3,4] This enabled us to investigate the biochemical alterations of plasma membrane molecules related to cold acclimation.

We will report here the results obtained so far in our laboratory concerned with lipid and protein changes in plasma membranes of various plants during cold acclimation.

II. LIPID CHANGES

Many studies have been carried out to find a causal relationship between lipid compositions and cold hardiness. The results are, however, variable and rather controversial from study to study.[5] A problem in these studies is that lipids are extracted from whole cells or, at best, isolated crude membranes such as other membrane systems and, therefore, the obtained results are not representative of those of plasma membranes per se.

As presented in Table 1, the lipid composition of plasma membranes isolated from orchard grass seedlings (*Dactylis glomerata* L.) is quite different from other endomembranes which were recovered in the lower phase of the phase partition system.[3] Plasma membranes are generally characterized as the high content of sterols. The phospholipids of plasma membranes are less unsaturated compared with other endomembranes and, therefore, the fluidity of the plasma membranes, as assessed by fluorescent polarization of embedded fluorophore, DPH, is usually lower than the other endomembranes such as ER, tonoplasts, and mitochondrial membranes, suggesting a relatively rigid membrane.[6] As presented in Table 2, in orchard grass seedlings, phospholipids of plasma membranes did not show any appreciable change in the fatty acid composition following cold acclimation, although phospholipids in other endomembranes increased the proportion of unsaturated fatty acids. Nearly the same results were obtained in winter rye seedlings[7] and Jerusalem artichoke tubers.[8] In extremely cold hardy woody species, (mulberry trees), however, plasma membrane phospholipids were observed to be significantly unsaturated during cold acclimation under natural conditions.[9] Therefore, the unsaturation of phospholipid fatty acids seems to be dependent on plant species, especially the degree of cold hardiness.

As presented in Table 3, the relative content of phospholipids, on the other hand, were generally observed to increase following cold acclimation in every plant. In some herbaceous plants such as orchard grass,[6] winter tye,[7] and Jerusalem artichoke tubers,[8] the relative content of free sterols increases following cold acclimation. Thus, it is assumed that a high lipid synthetic activity is a prerequisite for the development of cold hardiness of plants. As will be discussed in later sections, the observed lipid changes, however, are considered to be interdependent on the qualitative changes in plasma membrane proteins.

According to our recent studies on lipid compositions of plant plasma membranes and tonoplasts isolated from etiolated mung bean hypocotyls,[10] it was found that ceramide monohexoside (cerebroside) is one of the major lipid components in these membranes, especially in the tonoplasts which comprised about 17 mol% of the total tonoplast lipids. Differential scanning clarorimetric measurement of the glycolipid showed a high phase

Table 1
COMPARISON OF LIPID COMPOSITION AMONG DIFFERENT CELLULAR MEMBRANES ISOLATED FROM ORCHARD GRASS SEEDLINGS

Membrane fractions	Phospholipid (μmol/mg protein)	Free sterol	ASG[a]	SG[b]	Total sterols/ phospholipid (mol/mol)
Plasma membrane	1.05	1.15	0.07	0.09	1.25
ER + Golgi membranes	1.11	0.61	0.11	0.10	0.74
Mitochondrial membranes	0.44	0.02	nd	nd	0.05

[a] ASG — acylated sterylglucoside.
[b] SG — sterylglucoside.

Table 2
CHANGES IN FATTY ACID COMPOSITION OF PHOSPHOLIPIDS FOLLOWING COLD ACCLIMATION

Membrane fraction		Fatty acids (mol %) 16:0	18:0	18:1	18:2	18:3	20:0	Unsaturated/saturated
Plasma membrane	NA[b]	26.1	0.7	7.5	40.6	23.1	1.2	2.54
	ACC[c]	26.1	0.6	6.1	41.5	23.8	1.0	2.57
Endomembranes[a]	NA	23.9	0.8	9.6	34.4	30.2	0.7	2.92
	ACC	21.0	0.4	6.1	38.6	32.4	0.7	3.49

[a] Endomembranes — mixture of ER, mitrochondria, plastids Golgi which were recovered in the lower phase after phase partitioning.
[b] NA — not acclimated.
[c] ACC — acclimated.

Table 3
CHANGES IN RELATIVE CONTENT OF PHOSPOLIPID IN PLASMA MEMBRANE FOLLOWING COLD ACCLIMATION

Plant species	Phospholipid content (μmol/mg protein) NA	ACC
Morus bombysis	1.38	1.88
Dactylis glomerata	1.11	1.38
Secale cereale	1.04	1.26
Helianthus tuberosus	1.03	1.45

Table 4
CHANGES IN LIPID COMPOSITIONS OF PLASMA MEMBRANES ISOLATED FROM JERUSALEM ARTICHOKE TUBERS DURING COLD ACCLIMATION

	Sampling dates (μmol/mg protein)			
Lipids	Sept. 25	Oct. 22	Nov. 19	Dec. 16
Phospholipids	1.02 (39.3)	1.15 (41.5)	1.49 (44.5)	1.63 (44.6)
Sterols	1.18 (45.6)	1.29 (44.6)	1.49 (44.5)	1.66 (45.5)
Cerebroside	0.39 (15.1)	0.33 (11.9)	0.37 (11.0)	0.36 (9.9)
Total	2.59 (100)	2.77 (100)	3.35 (100)	3.65 (100)
Cold hardiness (°C)	−3	−4	−6	−8

Note: Values in parentheses mean molar percent.

transition temperature at 35°C, suggesting that the tonoplast lipid-bilayers may undergo a thermotropic phase transition under chilling temperatures.[25] In plasma membranes of Jerusalem artichoke tubers, it was also confirmed that the cerebroside was the major lipid component, comprising about 15 mol% in the total plasma membrane lipids in the tubers harvested in early autumn. As indicated in Table 4, the mole fraction of the cerebroside decreased significantly as cold hardiness increased. Recently, nearly the same results as ours have been reported by Lynch and Steponkus in winter rye seedlings.[11] The physical properties of the lipid are not determined as yet, however, the alterations of the mole fractions as well as the chemical structures, i.e., the fatty acid composition, may have a profound effect on the physical properties of the plasma membrane lipid bilayers and also the molecular interactions between lipids and proteins in the plasma membranes, because of the unique structure of the amide-linked fatty acids, i.e., hydroxylated long chain fatty acids. Based on these facts, further detailed studies are needed in the future to clarify the role of cerebroside in cold hardiness of plants.

III. PROTEIN CHANGES

Plasma membrans have important physiological functions such as cell wall biosynthesis,[12] active ion transports,[13] hormone action,[14] phytochrome responses,[15] disease resistance,[16] and stress responses, etc. Therefore, the membranes contained several types of proteins enrolled in these functions as well as structural proteins conferring to the membrane integrity. It is plausible that plasma membrane proteins are in a dynamic state, changing its structures and compositions related to the physiological conditions of plants including growth, development, and environmental responses. In fact, plasma membranes of mulberry tree bark cells showed a remarkable change in the protein compositions at the time of growth cessation from September to early October,[9] independently of development of cold hardiness. However, specific changes were determined in the protein composition during the cold acclimating process, which were closely related to cold hardiness.

After two-dimensional SDS-polyacrylamide gel electrophoresis of heat-denatured plasma membrane proteins of orchard grass[6] and winter rye seedlings,[7] it appeared that specific polypeptides were decomposed and newly synthesized following cold acclimation. Figure 1 represents the compositional changes of plasma membrane polypeptides in winter rye seedlings following cold acclimation. Thus, electrophoresis of SDS-solubilized and heat-denatured plasma membrane proteins may provide more detailed information about the alterations of polypeptide compositions during cold acclimation, which may suggest that

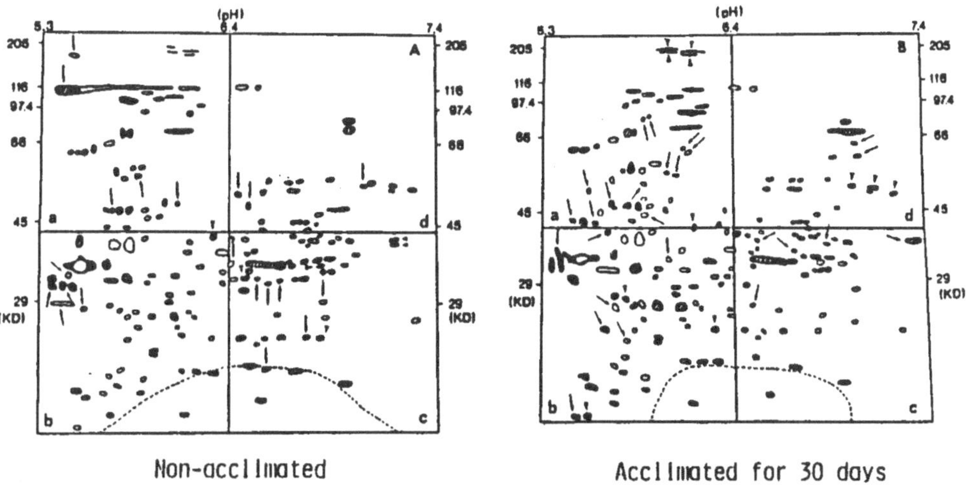

FIGURE 1. Tracings of two-dimensional SDS-polyacrylamide gel electrophoretograms of winter rye plasma membranes before and after cold acclimation. Disappeared or newly appeared polypeptide spots following cold acclimation are marked by arrows.

Table 5
CHANGES IN SOLUBILITY OF PLASMA MEMBRANE PROTEINS INTO TRITON X-114 DURING COLD ACCLIMATION

Sampling dates	% of solubilized membrane proteins
Sept. 25	51.8
Oct. 22	54.3
Nov. 19	60.7
Dec. 16	71.1

plasma membrane materials are actively metabolized. Nevertheless, it is sometimes rather difficult to reconcile the detailed compositional changes of polypeptides into the *in situ* structures and properties of plasma membranes in the natural state.

To get more insight into the structural features of plasma membrane proteins, we attempted to analyze the proteins under various conditions including a relatively mild solubilization of the membrane proteins with a neutral detergent and compared the results obtained with an ordinary SDS-solubilization method. Table 5 shows changes in the solubility of plasma membrane proteins of Jerusalem artichoke tubers into Triton X-114 during cold acclimation. As they increased the cold hardiness from −3 (September) to −8.8°C (December), the plasma membrane proteins became more soluble in the neutral detergent. This might indicate that plasma membrane proteins became more hydrophobic and/or were altered in the structures. Figure 2 indicates the SDS-polyacrylamide gel electrophoretograms of the Triton-soluble and insoluble proteins. There existed a distinct difference in the polypeptide compositions between the soluble and the insoluble protein fractions, suggesting that the detergent solubilized specific membrane proteins. Of the insoluble protein fraction, polypeptide bands of 52, 46, 30 and 25 kdaltons decreased the staining intensity, whereas the bands F-3 (one

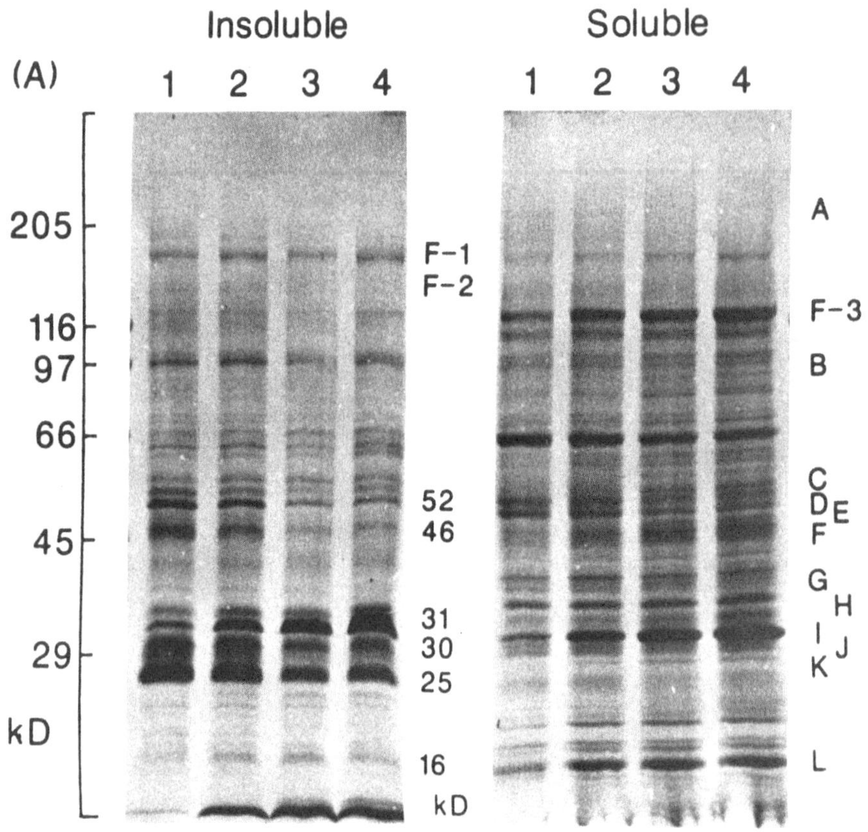

FIGURE 2. SDS-polyacrylamide gel electrophoretograms of Triton X-114 soluble and insoluble plasma membrane proteins. Plasma membranes were isolated from Jerusalem artichoke tubers at different season. The isolated plasma membranes were solubilized with 1% (w/v) Triton X-114 at 0°C. The insoluble and soluble protein fractions were subject to SDS-polyacrylamide gradient gel (5 to 15%) electrophoresis after solubilization in 2% SDS buffer solution contained 50 m*M* DTT with heating. Gels were stained with Coomassie blue. Lanes 1 to 4 correspond to sampling date, September 25, October 22, November 19, and December 16, respectively. F-1, F-2, and F-3 show the frost susceptible proteins. The polypeptide bands changed were marked by numerals (insoluble fraction) and letters (soluble fraction).

of the frost susceptible proteins as will be described later), B, C, F, H, I, J, and L in the soluble protein fraction increased the staining intensity during cold acclimation. There results may suggest that a replacement of less hydrophobic or relatively hydrophilic proteins by more hydrophobic proteins occurs in the plasma membranes during cold acclimation.

Triton X-114 is a polydisperse nonionic detergent and can be homogeneously dispersed at low temperature at 0°C, but reaches a clouding point above 20°C.[17] Above 20°C, a diluted solution of the detergent (1%, w/v), separates into two phases, a detergent-rich and a detergent-poor phase. During the phase separation, proteins are partitioned into both phases depending on the hydrophobicity. That is, integrated membrane proteins with a more amphiphilic or hydrophobic nature are preferentially partitioned into the detergent-rich phase, whereas less hydrophobic proteins are preferentially partitioned into the detergent-poor phase, although the partition may also be influenced by the molecular shapes and the conformations.[17] As indicated in Figure 3, polypeptide bands I (31 kdaltons) and J (30 kdaltons) were largely partitioned into the detergent-rich phase, suggesting the most hydrophobic polypeptides. These highly hydrophobic polypeptides showed a marked increase during cold acclimation. It is quite natural that a great deal of glycoproteins were insoluble in the detergent,

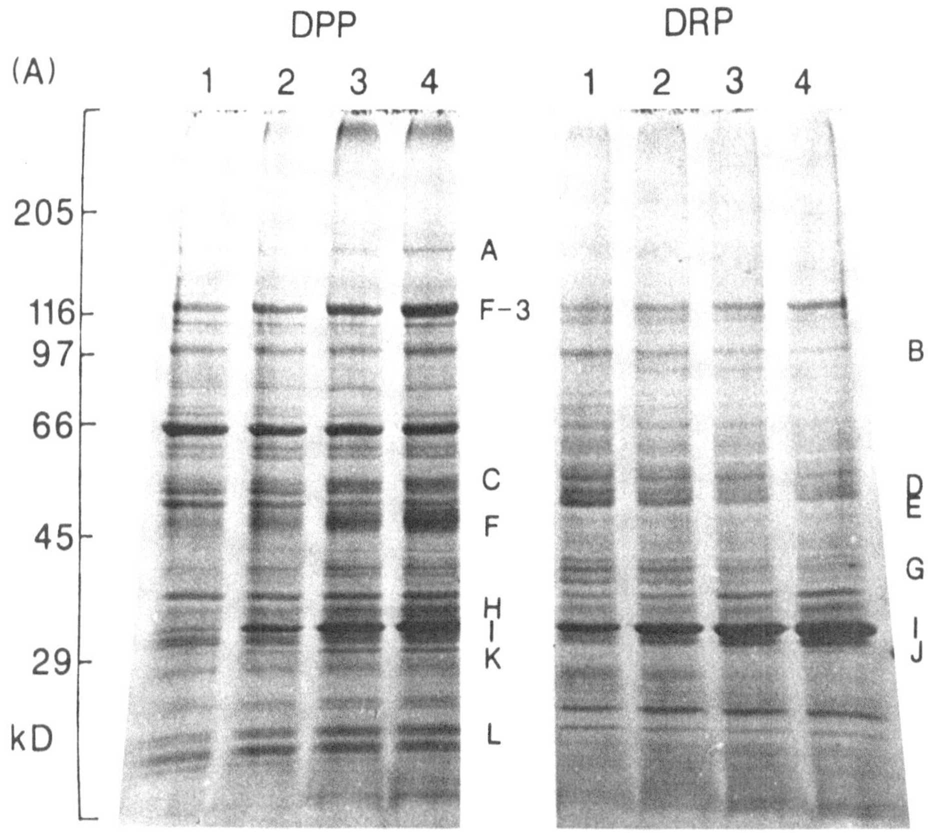

FIGURE 3. SDS-polyacrylamide gel (5 to 15%) electrophoretograms of Triton X-114 soluble plasma membrane proteins after phase-separation into detergent-poor (DPP) and detergent-rich (DRP) phases. Plasma membrane samples solubilized in 1% (w/v) Triton X-114 were phase-separated at 30°C according to the method of Bordier.[24] The DPP and DRP fractions were subject to a SDS-polyacrylamide gradient gel (5 to 15%) electrophoresis. Gels were stained by Coomassie blue. Lanes 1 to 4 refer to the legend in Figure 2. The bands changed the staining intensity during cold acclimation are indicated by letters on the side of the panels.

because of the hydrophilic nature of the sugar moieties attached to the proteins. However, it should be noticed that a significant amount of glycoproteins were solubilized and phase separated into the detergent-rich phase after the phase separation. This might indicate that in plant plasma membranes, the number of sugars bound to the proteins are unlikely limited in animal cells.

IV. ALTERATIONS OF PHYSICAL PROPERTIES

Vigh et al.[18] reported, using an ESR technique, that the plasma membranes in protoplasts isolated from winter wheat leaves became more fluid following cold acclimation. Upon using DHP-fluorescent polarization measurements of isolated plasma membranes no detectable change was observed in the membrane fluidity in most herbaceous plants, including orchard grass,[6] winter rye,[7] and Jerusalem artichoke.[8] This coincided with the minor change in the fatty acid unsaturation of phospholipids during cold acclimation. However, in extremely hardy woody species (mulberry tree), the plasma membrane became more fluidized during acclimation,[9] which agreed well with the increase in the degree of unsaturation of phospholipids. Thus, the fluidity change in plasma membrane is considered to be dependent on plant species, especially the degree of cold hardiness.

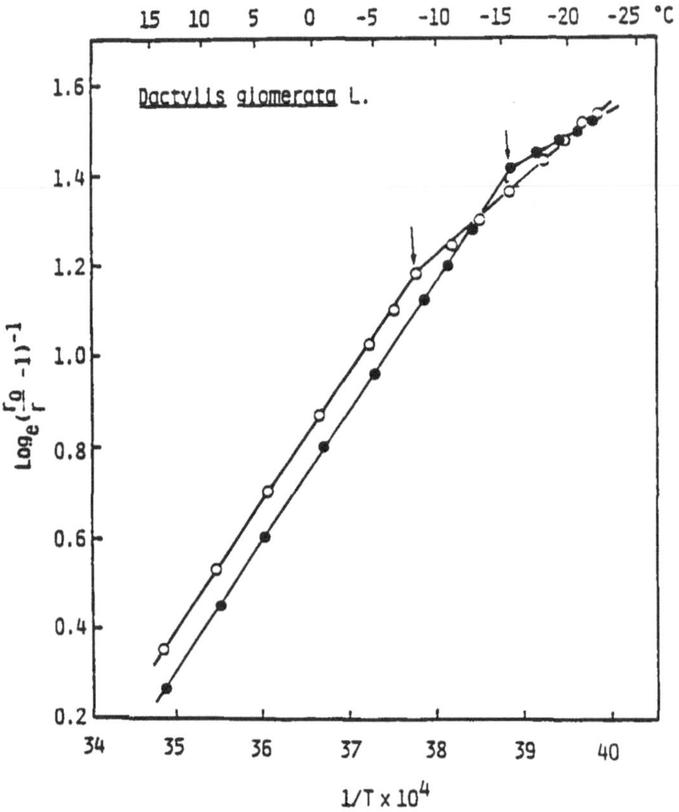

FIGURE 4. Arrhenius plots of fluorescence anisotropy parameters of DPH embedded into plasma membranes isolated from orchard grass seedlings in different seasons. ○ - ○, plasma membrane sample of early October (hardy to -7.5°C); ● - ●, plasma membrane sample of early December (hardy to -15°C). Arrows indicate inflection points.

When intact plant cells are frozen, the plasma membranes would experience more complex sorts of stresses and, therefore, it might be difficult to identify the specific factors responsible for the membrane damage. These factors may involve a physical effect of subfreezing temperatures per se,[19] a freeze-induced condensation of solutes,[20] a freeze-dehydration of plasma membrane molecules,[20] a freeze-induced cell shrinkage and molecular packing of plasma membrane molecules,[19] and/or a combination of these. From this point of view, isolated plasma membranes would be one of the useful materials to investigate on the molecular mechanism of freezing injury of plants. For this purpose, experiments were performed to gain information on the thermal properties of plasma membranes under subfreezing temperatures without freezing.

As indicated in Figure 4, and inflection was clearly detected in the Arrhenius plots of fluorescent anisotropy parameters of embedded fluorophore, DPH (1,6-diphenyl-1,3,5-hexatriene), in orchard grass plasma membranes. The temperatures for the inflections apparently corresponded with the temperatures that cause freezing injury. Because the measurements were performed with membrane samples suspended in a buffer solution containing 35% ethylene glycol to prevent freezing, the inflections may have reported a temperature-dependent alteration of the physical state of plasma membrane lipid-bilayers.

The fluorescence polarization is reported to be dependent not only on the rotational relaxation time of the fluorescence probe, DPH, but also on the degree of hinderance of the molecular rotation. Thus, the fluorescence anisotropy parameters obtained by steady-state fluorescent polarization measurements may represent the degree of static orientation constraint of the probe or the averaged freedom of molecular motion of lipid molecules and, thereby, a comparative index for the "fluidity" of the membranes.[20] Therefore, the inflection means, in one way, a phase transition of membrane lipids or more correctly an orientational constraint of the membrane molecules surrounding the probe. At any rate, it may be stated that some physical changes are produced in the plasma membrane molecules, proteins and lipids, and/or the molecular interaction at a certain subzero temperature. As a matter of fact, when the measurements were carried out with liposomes prepared from extracted lipids, the inflections occurred at much lower temperatures than those of intact plasma membranes.[21] This result may suggest that the inflection, i.e., the change of the physical state of plasma membrane molecules is highly dependent on membrane proteins, suggesting that proteins-lipid interaction is altered in the plasma membranes by subfreezing temperatures. Based on these facts, it appears that the observed molecular changes in plasma membranes during cold acclimation are reflective of the readjustment of the lipid-protein interaction so as to adapt to freezing.

Upon in vivo freezing of Jerusalem artichoke tubers to sublethal freezing temperatures, biochemical changes occurred on the plasma membranes.[22] Those were reduction of specific activity of ATPase and an increase of the DCCD-sensitivity, losses of specific proteins (designated as frost-susceptible proteins), and lipids such as sterols and phosphatidylethanolamine (PE). At the moment it is not easy to explain how these lipid and protein changes were brought about in the plasma membranes and how these changes are interdependent. However, the possibility cannot be ignored that the plasma membrane proteins may first undergo conformational alterations in a reversible or an irreversible manner as a consequence of lowering in temperatures and/or a freeze-dehydration. As a result, the specific interaction between lipids and proteins may be significantly disturbed and presumably result in segregation of sterols or PE-enriched lipid domains in the plane of plasma membranes, and also result in sensitization of the frost-susceptible proteins to a specific protease(s) and/or an activation of the enzyme(s) per se. As reported by Steponkus' group,[23] even if the lateral phase separation or the lamellar-to-hexagomal II phase transition occurs during freezing, it would be initiated only after the disturbance of lipid-protein interactions as a result of lowering in temperatures, whereby the lamellar-forming lipids such as phosphatidylethanolamine are arrested by the proteins.

V. SUMMARY

The results obtained by chemical analysis of plasma membranes suggest that they undergo several changes during the development of cold hardiness. Of these changes, marked increases of the relative contents of phospholipids and free streols on a milligram of protein basis, and a decrease of mole fraction of cerebroside in the total lipids were dominated. In parallel with these lipid changes, qualitative changes were also demonstrated to occur on the plasma membrane proteins, which may involve a replacement of less hydrophobic proteins by more hydrophobic proteins. Those changes in lipids and proteins may be a closely interdependent phenomenon, indicating a molecular reorganization of plasma membranes through an alteration of the hydrophobic interaction between lipids and proteins and probably an alteration of protein structure. The temperature-dependent alteration of the physical state of plasma membranes at a certain subfreezing temperature, as assessed by a fluorescent polarization measurement and the downward shift of the temperatures to cause the alteration following acclimation, could be interpreted by posturating an alteration of the lipid-protein interaction in the plasma membranes. From these points of view, the process of cold acclimation can be interpreted in terms of a molecular adaptation of plasma membranes to freezing stress.

For a better understanding of the mechanisms, we must accumulate more knowledge about the biochemical background of the molecular alterations of plasma membranes during cold acclimation and the biophysical features, including the structures of lipids and proteins, and, especially, the molecular interaction in the natural membranes. Also, in the future, we need more precise information as to how plasma membrane molecules, particularly the lipid-protein interaction and protein structures, are affected by freezing temperatures per se and/ or freeze-dehydration.

REFERENCES

1. **Levitt, J.,** *Responses of Plants to Environmental Stress,* Vol. 1, 2nd Ed., Academic Press, New York, 1980.
2. **Steponkus, P. L.,** Role of the plasma membrane in freezing injury and cold acclimation, *Annu. Rev. Plant Physiol.,* 35, 543, 1984.
3. **Yoshida, S., Uemura, M., Niki, T., Sakai, A., and Gusta, L. V.,** Partition of membrane particles in aqueous two-polymer phase system and its practical use for purification of plasma membranes from plant, *Plant Physiol.,* 72, 105, 1983.
4. **Uemura, M. and Yoshida, S.,** Isolation and identification of plasma membrane from light grown winter rye seedlings (*Secale cereale* L. cv Puma)., *Plant Physiol.,* 73, 586, 1983.
5. **Willemot, C.,** Chemical modification of lipids during frost hardening of herbaceous species, in *Low Temperature Stress in Crop Plants,* Lyons, J. M., Graham, D., and Raison, J. K., Eds., Academic Press, New York, 1979, 411.
6. **Yoshida, S. and Uemura, M.,** Protein and lipid compositions of isolated plasma membranes from orchard grass (*Dactylis glomerata* L.) and changes during cold acclimation, *Plant Physiol.,* 75, 31, 1984.
7. **Uemura, M. and Yoshida, S.,** Involvement of plasma membrane alterations in cold acclimation of winter rye seedlings *(Secale cereale* L. cv Puma), *Plant Physiol.,* 75, 818, 1984.
8. **Ishikawa, M. and Yoshida, S.,** Seasonal changes in plasma membranes and mitochondria isolated from Jerusalem artichoke tubers, possible relationship to cold hardiness, *Plant Cell Physiol.,* 26, 1331, 1985.
9. **Yoshida, S.,** Chemical and biophysical changes in the plasma membrane during cold acclimatin of mulberry bark cells *Morus bombysis* Koid. cv Goroji), *Plant Physiol.,* 76, 257, 1984.
10. **Yoshida, S. and Uemura, M.,** Lipid composition of plasma membranes and tonoplasts isolated from etiolated eedlings of mung bean *(Vigna radiata* L.), *Plant Physiol.,* 82, 807, 1986.
11. **Lynch, D. V. and Steponkus, P. S.,** Plasma membrane lipid alterations associated with cold acclimation of winter rye seedlings *(Secale cereale* L. cv Puma), *Plant Physiol.,* 83, 761, 1987.
12. **Montezinos, D. and Brown, R. M., Jr.,** Surface architecture of the plant cell: biogenesis of the cell wall with special emphasis on the role of the plasma membrane in cellulose biosynthesis, *J. Supramol. Struct.,* 5, 277, 1976.
13. **Sze, H.,** H^+-translocating ATPases: advances using membrane vesicles, *Annu. Rev. Plant Physiol.,* 36, 175, 1985.
14. **Stoddart, J. L. and Venis, M. A.** Molecular and subcellular aspects of hormone action, in *Hormonal Regulation of Development I, Enclyclopedia of Plant Physiology,* New Series Vol. 9, MacMillan, J., Ed., Springer-Verlag, Berlin, 1980, 455.
15. **Quail, P. H.,** Plant cell fractionation, *Annu. Rev. Plant Physiol.,* 30, 425, 1979.
16. **Strobel, G. A. and Hess, W. M.,** Evidence for the presence of toxin-binding protein on the plasma membranes of sugar cane cells, *Proc. Natl. Acad. Sci. U.S.A.,* 71, 1413, 1974.
17. **Alcaraz, G., Kinet, J-P., Kumar, N., Wank, S. A., and Metzger, H.,** Phase separation of the receptor for immunoglobin E and its subunit in Triton X-114, *J. Biol. Chem.,* 259, 14922, 1984.
18. **Vigh, L., Horvath, I., Horvath, L. I., Dudis, D., and Farkas, T.,** Protoplast plasmalemma fluidity of hardened wheats correlates with frost resistnce, *FEBS Lett.,* 107, 291, 1979.
19. **Pringle, M. J. and Chapman D.,** Biomembrane structure and effects of temperature, in *Effects of Low Temperatures on Biological Membranes,* Morris, G. J. and Clarke, A., Eds., Academic Press, New York, 1981, 21.
20. **Franks, F.,** Biophysics and biochemistry of low temperatures and freezing, in *Effects of Low Temperatures on Biological Membranes,* Morris, G. J. and Clarke, A., Eds., Academic Press, New York, 1981, 3.
21. **Yoshida, S.,** Studies on freezing injury of plant cells. I. Relation between thermotropic properties of isolated plasma membrane vesicles and freezing injury, *Plant Physiol.,* 75, 38, 1984.

22. **Uemura, M. and Yoshida, S.,** Studies on freezing injury of plant cells. II. Protein and lipid changes in the plasma membranes of Jerusalem artichoke tubers during a lethal freezing *in vivo*, *Plant Physiol.*, 80, 187, 1986.
23. **Gordon-Kamm, W. J. and Steponkus, P. L.,** Lammelar-to-hexagonal$_{II}$ phase transitions in the plasma membrane of isolated protoplasts after freeze-induced dehydration, *Proc. Natl. Acad. Sci. U.S.A.*, 81, 6373, 1984.
24. **Bordier, C.,** Phase separation of intergral membrane proteins in Triton X-114 solution, *J. Biol. Chem.*, 256, 1604, 1981.
25. **Yoshida, S., Nashio, K., Kenrick, J., and Orr, G.,** Thermotropic properties of lipids extracted from plasma membranes and tomoplast isolated from chilling-insensitive mungbean *(Vigna radiata* [L.] Wilczek) *Plant Cell Physiol.*, in press.

Chapter 4

DEVELOPMENT AT COLD-HARDENING TEMPERATURES: MEMBRANE ASSEMBLY AND ORGANIZATION

Norman P. A. Huner, Marianna Krol, John P. Williams, Ellen Maissan, and Z. Krupa

TABLE OF CONTENTS

I. Introduction 54

II. Low Temperature-Induced Alteration in Thylakoid Organization and Assembly 54

III. Specific Membrane Component Which Modulates LHCII Organization and Assembly 59

IV. The Generality of the Temperature-Induced Change in Fatty Acid Composition of PG 62

V. Conclusion 63

References 64

I. INTRODUCTION

Results of research into the mechanism of freezing injury in plants indicate, in general, that membranes, and in particular the plasma membrane, are the primary site of this injury.[1] Under natural conditions, exposure of plants to low temperatures over a prolonged period of time is mandatory for the initiation of a cold acclimation process and the subsequent establishment of a cold hardy state which imparts the property of freezing resistance to the plant. Thus, the primary focus of most biochemical research related to cold acclimation has been, first, to identify specific memrane components, in particular proteins and lipids, which have been altered in some way by a low temperature acclimation period and, second, to correlate these biochemical alterations with the capacity of the plant to withstand freezing temperature.[2-5] Huner and co-workers[6] have shown that leaves of winter rye *(Secale cereale* L. cv Puma) undergo distinct morphological, anatomical, and biochemical changes during growth and development at low, cold-hardening temperatures. Huner[6] suggests that cold acclimation, under natural conditions, is a direct consequence of plant growth and development at low, cold-hardening temperatures. Although chloroplast thylakoid membranes have been used as a model system for identifying the primary lesions induced by freezing,[7-9] we have employed the chloroplast thylakoid membranes to examine membrane assembly during growth and development at low temperature in order to understand the role(s) this process might play in plant acclimation to cold-hardening temperatures. Although the gross polypeptide, pigment, and lipid compositions are not drastically altered during development at cold-hardening temperatures, the supramolecular organization of rye thylakoid membranes is altered which is reflected in significant changes in thylakoid structure and function.[10-16] In this chapter, we present a summary of recent results which indicate that the observed changes in thylakoid membrane organization during growth and development at low temperature are, at least in part, due to an altered membrane assembly process. More importantly, we present evidence for a single specific change in the fatty acid composition of phosphatidyldiacyl-glycerol which we believe to be a key to understanding how low temperature modulates the supramolecular organization of an integral thylakoid membrane pigment-protein complex from winter rye.

II. LOW TEMPERATURE-INDUCED ALTERATION IN THYLAKOID ORGANIZATION AND ASSEMBLY

Chloroplast thylakoid membranes are highly specialized membranes consisting of pigments, proteins, and lipids whose specific organization imparts the capacity for efficient light trapping and energy conversion by way of two photochemical reactions which are connected in series via associated electron transport processes and energy coupling reactions involved in ATP synthesis. This energy transducing system is specifically organized into four major catalytic supramolecular protein complexes:[17,18] photosystem I (PSI) complex is the major Chl a pigment-protein complex containing the reaction center, P_{700}; photosystem II (PSII) is a Chl a pigment-protein complex which contains the reaction center, P_{680}, and the water-splitting system; cytochrome f/b_6 complex is a nonchlorophyll protein complex situated between PSI and PSII and catalyzes electron-proton exchange between these two photosystems; ATP synthetase complex (CF_0-CF_1) is a nonpigmented protein complex which is capable of utilizing the proton gradient generated by photosynthetic electron transport to synthesize ATP. In addition to these catalytic components, the thylakoid membrane also consists of two major noncatalytic supramolecular pigment-protein complexes: LCHII is the major light harvesting Chl a/b-protein complex closely associated with PSII. This pigment-protein complex can constitute as much as 60% of the total chlorophyll of the thylakoid membrane[18] and, when isolated, typically exhibits a Chl a/b ratio of 1.0 to 1.5; LCHI is the

Table 1
GENERAL COMPOSITION OF RH AND RNH THYLAKOIDS

Thylakoid Compositions	RNH	RH	Ref.
Chl a/b	2.99	2.96	12
Carotenoid/Chl (g/mℓ)	0.096	0.092	12
Protein/Chl	6.8	7.1	10
Chl/P_{700}	505	530	12

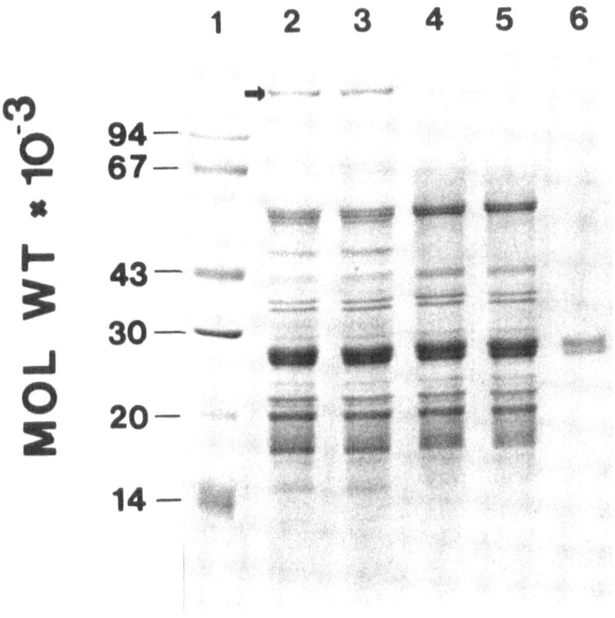

FIGURE 1. SDS gel illustrating RH and RNH thylakoid polypeptides. Lane 1, molecular weight standards; β-lactalbumin, 14,000; soybean trypsin inhibitor, 20,000; carbonic anhydrase 30,000; alcohol dehydrogenase, 43,000; ovalbumin, 67,000; bovine serum albumin, 67,000; phosphorylase b, 94,000. Lanes 2 and 3; RH and RNH thylakoids, nonheated. Arrow indicates polypeptides associated with PSI. Lanes 4 and 5; RH and RNH thylakoids heated for 60 sec at 100°C prior to electrophoresis. Lane 6, purified RNH LCHII.

light harvesting Chl a/b - protein complex associated with PSI typically exhibits a Chl a/b ratio of 3.4 to 4.0.[18] LCHII and LHCI act as antenna for PSII and PSI reaction centers respectively, thus affecting a more efficient gathering of available light energy which is subsequently transferred to the respective reaction center. In addition, LHCII has also been shown to involved in the modulation of energy distribution between PSI and PSII.[17] All of the protein complexes discussed above are intrinsic components of the chloroplast thylakoid membrane except for the CF_1 portion of the ATP synthetase complex which is an extrinsic component on the stromal side of the thylakoid membrane.[18] In addition, all chlorophyll within the thylakoid membrane is specifically associated with the pigment-protein complexes discussed above. There is no free chlorophyll within the thylakoid membrane.[19]

Although the general pigment and protein content (Table 1) as well as the polypeptide complement (Figure 1) of rye thylakoids developed at 5°C are not significantly different from thylakoids developed at 20°C (RNH), thylakoids developed at cold-hardening tem-

Table 2
COMPARISON OF SOME SPECIFIC STRUCTURAL AND FUNCTIONAL PROPERTIES OF RH AND RNH THYLAKOIDS

Properties	RNH Thylakoids	RH Thylakoids	Ref.
Osmotic concentration required to prevent H^+ leakage	≤0.4 *M*	≥0.8 *M*	15
Particle size distribution EF_s fracture face	Bimodal (105A and 150A)	Unimodal (136A)	12
Granal stacking	Greater	Reduced	12
Chl extractability with non-ionic detergents	Greater	Reduced	16
Sensitivity of LHCII to precipitation with MG^{2+}/K^+	More	Less	20
Light-saturated rates for PSI	400	800	11
F_{685}/F_{742}	1.19	1.42	14
Cooperativity between LHCII-PSII units	Greater	Reduced	14

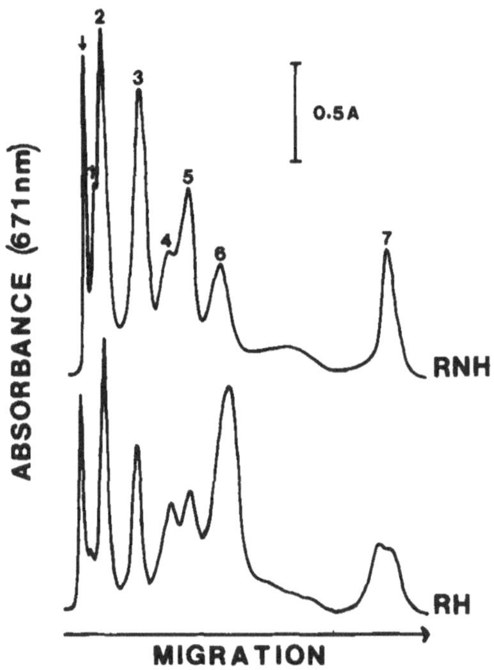

FIGURE 2. Densitometric gel scans of chlorophyll-protein complexes from RH and RNH thylakoids. Small arrow indicates top of the gel. Peaks 1 and 2; CPla and CPl which are the chlorophyll a-protein complexes associated with PSI. Peaks 3,4, and 6; $LHCP_1$, $LHCP_2$, and $LHCP_3$, respectively, which are the oligomeric, dimeric, and monomeric forms, respectively, of the chlorophyll a/b light harvesting-protein complex associated with PSII (LHCII). Peak 5; CPa which is the chlorophyll a protein complex containing the reaction center for PSII. Peak 7; free pigment.

peratures (5°C) did exhibit significant differences in both structural and functional properties from thylakoids developed at nonhardening, control temperatures (20°C) (Table 2). The most striking structural difference between RH and RNH thylakoids is related to the organization of LHCII. The various chlorophyll-protein complexes of rye thylakoid membranes can be visualized by careful dissolution of the membrane with detergents and subsequent separation by SDS gel electrophoresis.[21] A comparison of the Chl-protein complexes of RH and RNH thylakoids from mature plants after such a separation is illustrated by the gel scan in Figure 2. Each green band on the gel was identified by its characteristic absorption spectrum. Peaks 1 and 2 are the major Chl a-protein complexes associated with PSI; peak 5 is the Chl a-protein complex associated with PSII; peaks 3, 4, and 6 are the major Chl a/b-protein complexes associated with LHCII. Peak 3 is oligomeric LHCII, peak 4 is dimeric LHCII and peak 6 is monomeric LHCII. Peak 7 represents free pigment which typically constitutes 5 to 10% of the total chlorophyll. Generally, the ratio of oligomeric LHCII: monomeric LHCII in thylakoids developed at 20°C (RNH) is at least double that observed in thylakoids developed at 5°C (RH).[21] Huner and co-workers have shown that this particular property of rye thylakoids is directly dependent upon the developmental temperature between 0 and 20°C.[21] These in vitro results are consistent with the following *in situ* results. The particles observed on the EF_s fracture face of thylakoid membranes have been shown to be associated with LHCII-PSII units.[17] Freeze fracture of RH and RNH thylakoids have indicated a shift from the typical bimodal particle size distribution (105A and 150A) to the unique unimodal particle size distribution (136A) (Table 2). In addition, the change in the 77°K fluorescence excitation at 685 nm (F_{685}) relative to that at 742 nm (F_{742}) as well as the change in cooperativity between LHCII-PSII units observed between RH and RNH thylakoids (Table 2) are consistent with a shift from predominantly oligomeric LHCII to monomeric LHCII or some intermediate form of LHCII.[22] Thus, *in situ* and in vitro results indicate that LHCII undergoes significant changes in organization when rye plants develop at low, cold-hardening temperatures.

The biogenesis of the photosynthetic apparatus in monocotyledonous plants can be conveniently studied during greening of etiolated seedlings.[23] We examined, in detail, the assembly of LHCII during the light-induced transformation of etioplasts to mature chloroplasts at either 5 or 20°C.[30] Figure 3 illustrates the chlorophyll-protein complexes separated on SDS gels as function of greening time at 20°C. between 0 and 24 hr of continuous light, monomeric LHCII is the predominant form of the light harvesting complex (Figure 3, peak 5). However, upon further greening at 20°C, oligomeric LHCII gradually becomes the predominant form of the light harvesting complex (Figure 3, peak 2). The results in Figure 4 indicate that the ratio of oligomeric LHCII: monomeric LHCII ($LHCP_1$:$LHCP_3$) gradually increases from an initial value of about 0.3 to a final value of about 2.5 in mature leaves (M). In contrast, the ratio of oligomeric LHCII: monomeric LHCII remains very low (0.30) during greening at 5°C. Thus, thylakoid biogenesis at 20°C results in the assembly of monomeric LHCII units into oligomeric units whereas the assembly of oligomeric LHCII is prevented, in some way, during thylakoid biogenesis at 5°C. Fluorescence excitation spectra (77°k) of rye thylakoids confirmed that the organization and assembly of LCHII during greening at 20°C differed significantly from that observed during greening at 5°C.

The assembly of the photosynthetic apparatus during greening of etiolated rye seedlings under continuous light was also studied by monitoring the development of various photochemical activities.[24] it is generally accepted that the greening of etioplats results in a sequential appearance of electron transport activities with PSI developing prior to PSII activity.[25] Chloroplast biogenesis at low temperature does not alter the sequence for the development of PSI relative to PSII activity but, rather, specifically alters the development of PSI activity relative to chlorophyll synthesis. The pattern for the development of PSI associated electron transport at 20°C is *antiparallel* to the accumulation of chlorophyll, that is, within the first 3 hr of greening PSI activity is maximum while the rate of chlorophyll synthesis is at a minimum. Subsequently, PSI activity decreases exponentially to a steady-state level during the time when the rate of chlorophyll synthesis is maximum. This pattern for the development of PSI activity is consistent with the results published for other greening systems.[25] In contrast, the pattern for the development of PSI activity during greening at cold-hardening temperatures is quite unique; the appearance of PSI activity is *parallel* to chlorophyll synthesis throughout development under continuous light. In addition, the steady-state PSI activity after greening at 5°C is double that oberved for PSI after greening at 20°C.

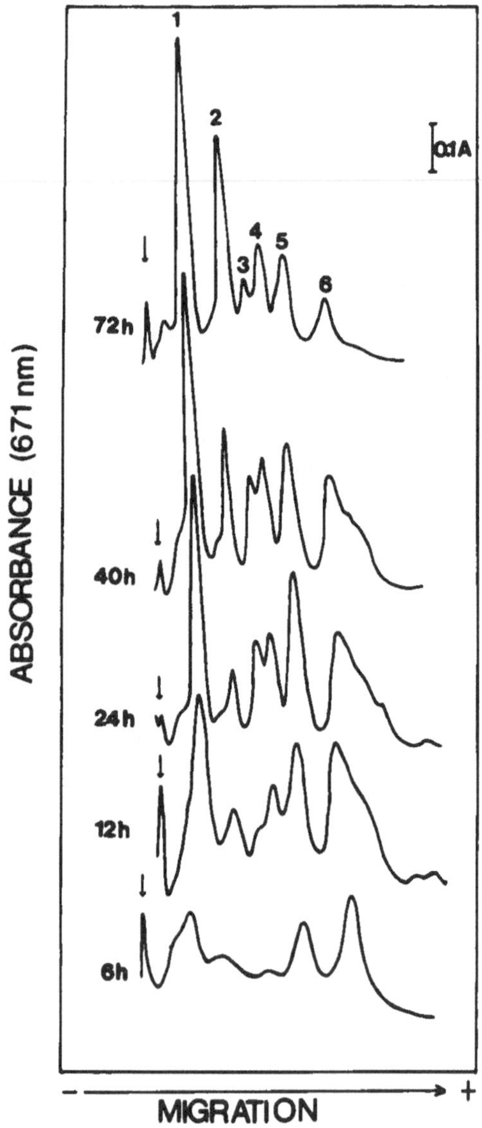

FIGURE 3. Densitommetric scans of chlorophyll-protein complexes associated with RNH thylakoids during greening at 20°C. Small arrow indicates the top of the gel. Peak 1, CPI, Peak 2, 3, and 5 are oligomeric, dimeric, and monomeric forms of LHCII, respectively. Peak 4, CPa. Peak 6, free pigment. Numbers to the left of each scan represent the numer of hours that etiolated rye seedlings were exposed to continuous light at 20°C prior to thylakoid isolation.

This is consistent with results obtained with mature plants (Table 2). The developmental patterns observed for PSI could be reversed by simply changing the temperature to which the etiolated rye seedlings were exposed during greening. For example, etiolated rye seedlings developed at 5°C in the dark, exhibit the antiparallel pattern and not the parallel pattern when exposed to continuous light at 20°C. Based on the observed developmental patterns for LHCII and PSI activity at 5 and 20°C, we suggest that development at cold-hardening temperatures induces an alteration in the pattern for thylakoid membrane assembly, which, as a consequence, leads to altered membrane stability and organization.

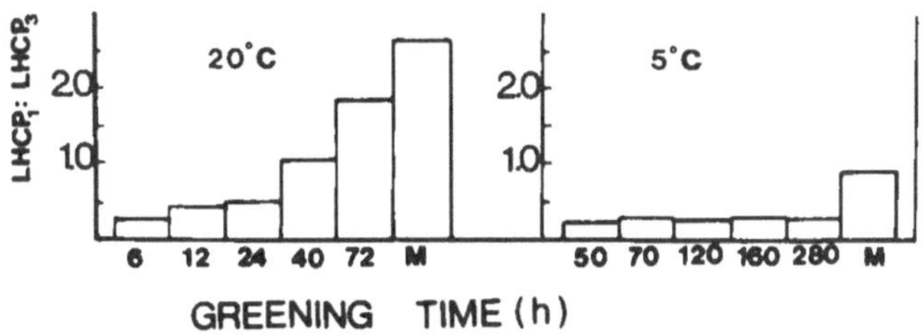

FIGURE 4. The effect of greening on the ratio of oligomeric LHCII: monomeric LHCII. $LHCP_1$, oligomeric LHCII; $LHCP_3$ monomeric LHCII. Etiolated rye seedlings were exposed to continuous light at either 20 or 5°C and thylakoids isolated at various times. Chlorophyll-protein complexes were separated by gel electrophoresis[21] as in Figure 3. The gels were scanned at 671 nm and the relative peak areas determined from which the $LHCP_1$:$LHCP_3$ was calculated. Etiolated rye seedlings were developed at either 20° or 5°C in the dark.

III. SPECIFIC MEMBRANE COMPONENT WITH MODULATES LHCII ORGANIZATION AND ASSEMBLY

Phosphatidyldiacylglycerol (PG) is the major phospholipid present in thylakoid membranes. When present in higher plant chloroplasts this, and only this phospholipid is characterized by the presence of an unusual fatty acid, *trans*-Δ^3-hexadecanoic acid (*trans* 16:1).[26] When present in PG *trans* 16:1 is specifically esterified to carbon-2 only. In our intensive search for alteration(s) in various thylakoid membrane components (pigments proteins, and lipids) which could account for the observed reorganization of RH thylakoids, we have recently reported a single, very specific change in the fatty acid composition of PG. Although no change in total lipid content was observed between RH and RNH thylakoids,[21] growth at cold-hardening temperatures did result in a specific 72% decrease in the *trans*-16:1 content associated with PG with a concomitant increase in the hexadecanoic acid (16:0) content of PG. The ratio of *trans* 16:1/16:0 in PG was 1.1 for RNH thylakoids but only 0.2 for RH thylakoids (Table 3).. The decrease in the *trans* 16:1 content of rye thylakoid PG was directly proportional to the decrease in growth temperature between 20 and 0°C. More importantly, the decrease in rye thylakoid *trans*-16:1 content was directly proportional to the decrease in the ratio of oligomeric LHCII: monomeric LHCII during growth at temperatures between 20 and 0°C.[21] This linear relationship between the *trans*-16:1 content and supramolecular organization of rye LCHII exhibited an r^2 value of 0.93.[21] This is consistent with the work of Dubacq and Tremolieres.[27] Developmental temperatures between 20 and 5°C did not affect significantly the 18:3 content of PG nor did it affect the fatty acid composition of the other major lipid classes of rye thylakoid membranes: monogalactosyldiacylglyceride (MGDG), digalactosyldiacylglycerol (DGDG), and sulfoquinovosyldiacylglyceride (SQDG). This specific change in the fatty acid composition of PG did not affect thylakoid membrane fluidity as measured by differential scanning calorimetry, fluorescence polarization of diphenylhexatriene, and electron spin resonance of 16-doxylstearic acid.[21]

The synthesis of *trans*-16:1 is light dependent.[26] We extended our observations with respect to the temperature-dependent change in the fatty acid composition of PG by examining the lipid composition of thylakoid membranes as a function of greening at 20 and 5°C. The results summarized in Figure 5 indicate that the kinetics for the synthesis of *trans*-16:1 and 16:0 in PG during development under continuous light at 20°C are quite distinct from that observed during development at 5°C. The ratio of *trans*-16:1/16:0 in PG increased immediately from about 0 to a maximum of 1.8 upon exposure of RNH etiolated seedlings to continuous light for 24 hr at 20°C. In contrast, the ratio of *trans*-16:1/16:0 exhibited an

Table 3
COMPARISON OF THE RATIO OF *TRANS*-16:1/16:0 IN PG OF RH AND RNH PREPARATIONS

Preparations	RNH	RH	Ref.
Whole leaf extracts	0.86	0.15	21
Isolated thylakoids	1.06	0.18	21
Purified LHCII	1.48	0.32	20

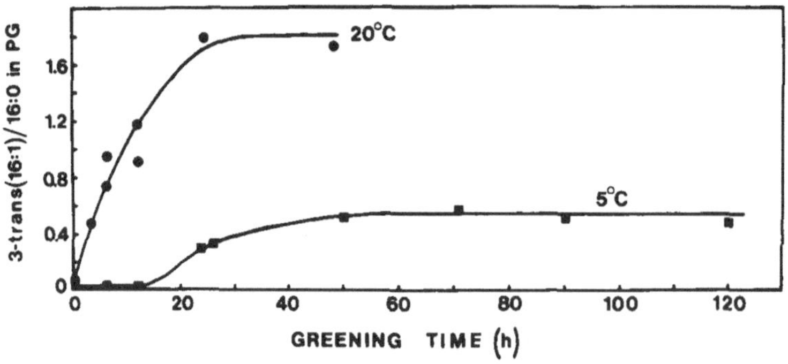

FIGURE 5. The effect of greening on the fatty acid composition of PG. Lipid and fatty acid analyses were performed on etiolated rye seedlings[21] as a function of greening time at either 20° or 5°C. The ratio of *trans*-16:1/16:0 in PG was calculated. Etiolated rye seedlings were developed at either 20° or 5°C in the dark.

intial lag of 16 to 20 hr upon exposure of RH etiolated seedlings to continuous light at 5°C and, subsequently, this ratio increased to a maximum of only 0.6 after 50 to 60 hr at 5°C. Greening temperature did not exhibit a differential effect on the 18:3 content of PG.[30] Comparison of the results illustrated in Figure 5 with those in Figure 4 indicates that during thylakoid biogenesis at 20°C, a maximum ratio of *trans*-16:1/16:0 is attained after 24 hr whereas maximum ratio of oligomeric LHCII: monomeric LHCII is attained after this time. Thus, maximization of *trans*-16:1 content in PG precedes the assembly of monomeric LHCII into predominantly oligomeric LHCII. In contrast, the ratio of *trans*-16:1/16:0 in PG during thylakoid biogenesis at 5°C remains comparatively low, such that monomeric or some intermediate form of LHCII predominates.

Such a strong correlation between the supramolecular organization of rye LHCII and the fatty acid composition of PG suggested to us that PG may be an integral structural component of rye LHCII. Lipid analysis of purified LHCII from both RH and RNH thylakoids[20] (see Figure 1, lane 6) indicated a fivefold enrichment of PG relative to the thylakoid membrane bilayer lipids MGDG and DGDG. This enrichment was specific for PG. Although the PG content of purified RNH and RH LHCII was not significantly different, the ratio of *trans*-16:1/16:0 was fivefold higher in RNH than in RH LCHII (Table 3). In addition, the ratio of oligomeric LHCII: monomeric LHCII of purified RNH LHCII was 60% higher than that observed for purified RH LHCII. Thus, the fatty acid and structural analyses of purified RH and RNH LHCII are consistent with those results for whole thylakoids. Furthermore, the observed specific enrichment of purified rye LHCII with respect to PG indicates that this phosopholipid is indeed an important structural component.

However, can this specific difference in the fatty acid composition of PG completely account for the observed difference in the supramolecular organization of RH and RNH

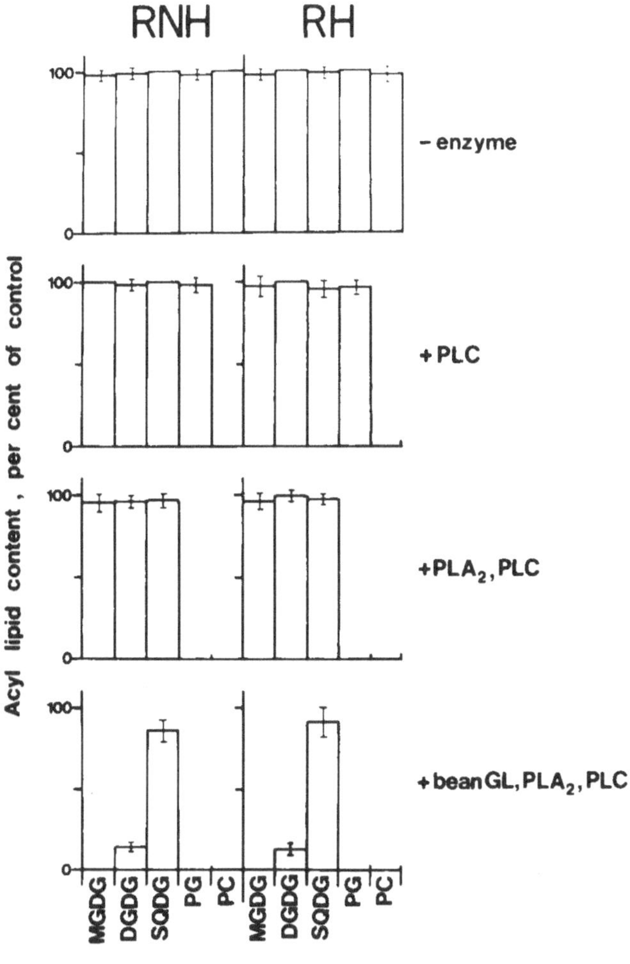

FIGURE 6. The effect of phospholipase treatment on the lipid composition of purified RH and RNH LHCII. The thylakoid lipids and the phospholipases employed have been described in the text. The methodology has been described in detail elsewhere.[20] All incubaitons were at 25°C for 90 min in the dark.

LHCII? To answer this question, we subjected purified RH and RNH LHCII to treatment with a series of phospholipases (PL) to hydrolyze different classes of lipids (Figure 6). Controls indicated that the incubation conditions did not significantly affect the lipid content of RH and RNH LHCII. Phospholipase C (PLC) treatment specifically hydrolyzed essentially all of the phosphatidylcholine (PC) and left 98% of the other major thylakoid lipids (MGDG, DGDG, SQDG, and PG) associated with LHCII. Incubation with phospholipase A_2 (PLA_2) and PLC completely removed all of the PG and PC but did not significantly affect the content of MGDG, DGDG or SQDG. After treatment of RH and RNH LHCII with bean galactolipase (GL) + PLA_2 + PLC, 85 to 90% of the SQDG and only minor amounts of DGDG (15%) remained associated with LHCII. Immediately after phospholipase treatment, the RH and RNH LHCII were subjected to SDS-PAGE to examine the relative distributions of oligomeric LHCII and monomeric LHCII. The results summarized in Table 4 clearly indicate that the removal of PG caused a significant decrease in the ratio of oligomeric LHCII: monomeric LHCII. Further treatment with bean galactolipase to remove MGDG and DGDG had very little effect on this ratio. More importantly, upon removal of PG, both RH

Table 4
THE EFFECT OF PHOSPHOLIPASE TREATMENT ON THE ORGANIZATION OF PURIFIED RH AND RNH LHCII

	Oligomeric LHCII:Monomeric LHCII	
Treatments	RNH LHCII	RH LHCII
Control	2.00	1.25
PLC	1.90	1.20
PLA_2 + PLC	0.65	0.65
Bean GL + PLA_2 + PLC	0.70	0.71

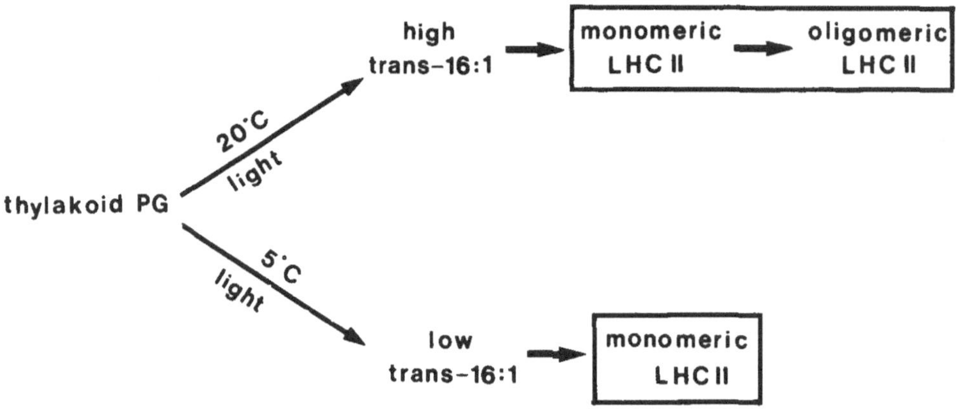

FIGURE 7. Summary of the effect of greening at 20° and 5°C on the *trans*-16:1 content of rye thylakoid PG and its subsequent effect on LHCII organization.

and RNH LHCII preparation exhibited the same ratio. Thus, we conclude that the difference in the organization of RH and RNH LHCII must be due to the fatty acid composition of PG.

Figure 7 illustrates a possible mechanism for the effect of cold-hardening temperature on the assembly and organization of rye LHCII. We suggest that development at cold-hardening temperatures modulates the assembly and the supramolecular organization of rye LHCII by specifically altering the *trans*-16:1 content of PG such that the oligomeric form predominates at the higher growth temperature whereas the monomeric or some intermediate form of LCHII predominates at the lower growth temperature. The physiological significance of this temperature-dependent change in the organization of rye LHCII is presently under intensive investigation.

IV. THE GENERALITY OF THE TEMPERATURE-INDUCED CHANGE IN FATTY ACID COMPOSITION OF PG

Preliminary results summarized in Table 5 indicate that all cold-tolerant species *do not* exhibit a developmental temperature-induced change in the fatty acid composition of chloroplast PG. All cold-tolerant monocotyledonous plant species that we have examined to date do exhibit the low temperature-induced decrease in the *trans*-16:1 content of PG. In addition, these plant species exhibited a low temperature-induced change in the ratio of oligomeric LHCII: monomeric LHCII. In contrast, the cold-tolerant dicotyledonous plant species examined to date exhibited neither a low temperature-induced decrease in the *trans*-16:1 content

Table 5
EFFECT OF GROWTH TEMPERATURE ON THE FATTY ACID COMPOSITION OF PG IN SOME COLD-TOLERANT PLANT SPECIES

	trans-16:1/16:0 in PG	
Plant species	**20°C**	**5°C**
Winter rye	1.0	0.21
Winter wheat	1.6	0.63
Broad bean	1.5	1.8
Spinach	1.6	1.5
Pea	1.3	1.4

of PG nor any change in the apparent organization of LHCII as measured by the ratio of oligomer: monomer. Thus, we emphasize that the proposed role for *trans*-16:1 of PG in modulating the organization of LHCII may become evident only under certain environmental conditions and, more importantly, this role may manifest itself to varying degrees in different plant species.

A role for the *trans*-16:1 in PG in the oligomerization of LHCII was first proposed by Dubacq and Tremolierses[27] and was recently refuted by Sommerville and co-workers.[28,29] We believe our work has provided some important insights which can, in part, account for these apparent discrepancies. We suggest that the original, proposed role[27] for PG must be altered in the following ways: (1) PG is not an absolute requirement for oligomerization of LHCII but, rather, enhances the stability of the oligomerized state of LHCII *in situ*.[20,21] (2) The dependence of LHCII oligomerization on the fatty composition of PG is species specific and (3) the role which PG plays in stabilizing the organization of LHCII appears to be environmentally modulated.

V. CONCLUSION

We have presented a summary of our recent data which indicate, for the first time, that growth and development of rye at low, cold-hardening temperatures results in a very specific change in the fatty acid composition of PG. This phospholipid appears to be specifically associated with an integral membrane protein complex of rye thylakoids and is directly involved in modulating its assembly and organization. That the biogenesis of rye thylakoid membranes as a whole is affected by development at cold-hardening temperature is further supported by the change observed for the assembly of PSI. We conclude that it is the altered membrane assembly process and not gross changes in membrane composition that has altered the general structural organization and functional properties rye thylakoid membranes. Clearly, not all plant species react in the same manner to a change in developmental temperature. We suggest that a greater focus on the effects of temperature on membrane biogenesis, membrane assembly processes, and lipid-protein interactions will lead to a greater understanding of the effects of low temperature acclimation on membrane structure and function. The chloroplast thylakoid membrane is an ideal system for such studies.

REFERENCES

1. **Steponkus, P. L.,** Role of the plasma membrane in freezing injury and cold acclimination, *Annu. Rev. Plant Physiol.*, 35,543, 1984.
2. **Uemura, M. and Yoshida, S.,** Involvement of plasma membrane alteration during cold acclimation of winter rye seedlings *(Secale cereale* L. cv Puma), *Plant Physiol.*, 75, 818, 1984.
3. **Yoshida, S.,** Chemical and biophysical changes in the plasma membrane during cold acclimation of mulberry bark cells *(Morus bombysis* Koidz. cv Goroji), *Plant Physiol.*, 76, 25, 1984.
4. **Yoshida, S. and Uemura, M.,** Protein and lipid composition of isolated plasma membranes from orchard grass *(Dactylis glomerata* L.) and changes during cold acclimation, *Plant Physiol.*, 75, 31, 1984.
5. **Heber, U.,** Freezing injury in relation to loss of enzyme activities and protection against freezing, *Cryobiology*, 5, 1988, 1968.
6. **Huner, N. P. A.,** Morphological, anatomical, and molecular consequences of growth and development at low temperature in *Secale cereale* L. cv Puma, *Am. J. Bot.*, 72, 1290, 1985.
7. **Garber, M. P. and Steponkus, P. L.,** Alterations in chloroplast thylakoids during cold acclimation, *Plant Physiol.*, 57, 673, 1976.
8. **Garber, M. L. and Steponkus, P. L.,** Alterations in chloroplasts thylakoids during and *in vivo* freeze-thaw cycle, *Plant Physiol.*, 57, 681, 1976.
9. **Klosson, R. J. and Krause, G. H.,** Freezing injury in cold acclimated and unhardened spinach leaves. I. Photosynthetic reaction of thylakoids isolated from frost-damaged leaves, *Planta*, 151, 339, 1981.
10. **Griffith, M., Brown, G. N., and Huner, N. P. A.,** Structural changes in thylakoid proteins during cold acclimation and freezing of winter rye (*Secale cereale* L. cv Puma), *Plant Physiol.*, 70, 418, 1982.
11. **Huner, N. P. A.,** Acclimation of winter rye to cold-hardening temperatures results in an increased capacity for photosynthetic electron transport, *Can. J. Bot.*, 63, 506, 1985.
12. **Huner, N. P. A., Elfman, B., Krol, M., and McIntish, A.,** Growth and development at cold-hardening temperatures. Chloroplast ultrastructure, pigment content and composition, *Can. J. Bot.*, 62, 53, 1984.
13. **Elfman, B., Huner, N. P. A., Griffith, M., Krol, M., Hopkins, W. G., and Hayden, D. B.,** Growth and development at cold-hardening temperatures. Chlorophyll-protein complexes and thylakoid membrane polypeptides, *Can. J. Bot.*, 62, 61, 1984.
14. **Griffith, M., Huner, N. P. A., and Kyle, D. J.,** Flourescence properties indicate that photosystem II reaction centres and light harvesting complex are modified by low temperature growth in winter rye, *Plant Physiol.*, 76, 381, 1984.
15. **Huner, N. P. A. and Hopkins, W. G.,** Growth and development of winter rye at cold-hardening temperatures results in thylakoid membranes with increased sensitivity to low osmoticum, *Physiol. Plant.*, 64, 468, 1985.
16. **Griffith, M., Huner, N. P. A., and Hayden, D. B.,** Low temperature development of winter rye leaves alters the detergent solubilization of thylakoid membranes, *Plant Physiol.*, 81, 471, 1986.
17. **Kaplan, S. and Arntzen, C. J.,** Photosynthetic membrane structure and function, in *Photosynthesis Energy Conversion by Plants and Bacteria*, Vol. 1, Govindjee, Ed., Academic Press, New York, 1982, chap. 3.
18. **Barber, J.,** Thylakoid membrane structure and organization of electron transport components, in *Photosynthetic Mechanisms and the Environment*, Barber, J. and Baker, N. R., Eds., Elsevier, Amsterdam, 1985, chap. 3.
19. **Baker, N. R. and Markwell, J. P.** Pigment-protein complexes and their interaction, in *Photosynthetic mechanisms and the Environment*, Barber, J. and Baker, N. R., Eds., Elsevier, Amsterdam, 1985, chap. 2.
20. **Krupa, Z., Huner, N. P. A., Williams, J. P., Maissan, E., and James, D. R.,** Development of cold-hardening temperatures. The structure and composition of purified rye light harvesting complex II, *Plant Physiol.*, 84, 19, 1987.
21. **Huner, N. P. A., Krol, M., Williams, J. P., Maissan, E., Low, P. S., Roberts, D., and Thompson, J. E.,** Low temperature development induces a specific decrease in trans-Δ^3-hexadecanoic acid content which influences LHCII organization, *Plant Physiol.*, 84, 12, 1987.
22. **Argyroudi-Akoyunoglou, J. H. and Akoyunoglou, G.,** Supramolecular structure of chlorophyll-protein complexes in relation to the chlorophyll a fluorescence of chloroplasts at room temperature or liquid nitrogen temperature, *Arch. Biochem. Biophys.*, 227, 249, 1983.
23. **Leech, R. M.,** Chloroplast development in angiosperms: current knowledge and future prospects, in *Chloroplast Biogenesis*, Baker, N. R. and Barber, J., Eds., Elsevier, Amsterdam, 1984, chap. 1.
24. **Krol, M., Huner, N. P. A., and McIntosh, A.,** Choloroplast biogenesis at cold-hardening temperatures. Development of photosystem I and photosystem II activities in relation to pigment accumulation, *Photosyn. Res.*, 14, 97, 1987.
25. **Baker, N. R.,** Development of chloroplast photochemical functions, in *Chloroplast biogenesis*, Baker, N. R. and Barber, J., Eds., Elsevier, Amsterdam, 1984, chap. 5.

26. **Harwood, J. L.** Plant acyl lipids: structure, distribution and analysis, in *The Biochemistry of Plants*, Vol. 4., Stumpf, P. K. and Conn. E. E., Eds., Academic Press, New York, 1980, chap. 1.
27. **Dubacq, J. P. and Tremolieres, A.,** Occurrence and function of phosphatidyl glycerol containing 3-trans-hexadecenoic acid in photosynthtic lamellae, *Physiol. Veg.*, 21, 293, 1983.
28. **McCourt, P., Browse, J., Watson, J., Arntzen, C. J., and Sommerville, C.,** Analysis of photosynthetic antenna funtion in a mutant of *Arabidopsis thaliana* (L) lacking trans-hexadecenoic acid, *Plant Physiol.*, 78, 853, 1985.
29. **Browse, J., McCourt, P., and Sommerville, C.** A mutant of *Arabidopsis* lacking a chloroplast-specific lipid, *Science* , 227, 763, 1985.
30. **Krol, M., Huner, N. P. A., Williams, J. P., and Maissan, E.,** *Photosynth. Res.*, 15, 115, 1988.

Chapter 5

THE ROLE OF PHENOLIC COMPOUNDS IN PLANT STRESS RESPONSES

Linda Chalker-Scott and L. H. Fuchigami

TABLE OF CONTENTS

I. Introduction 68

II. Phenolic Biosynthesis 68
- A. Enzymatic Regulation 68
- B. Cell-Wall Phenolics 70

III. Functions of Phenolic Compounds 71
- A. Growth Regulation 71
- B. Disease Resistance 71
- C. General Stress Responses 72
- D. Interactions With Ethylene 73
- E. Cold Stress Response 73

IV. Conclusions 75

References 76

I. INTRODUCTION

Plant phenolic compounds belong to a poorly defined group of products called secondary metabolites. Originally, secondary products were those regarded as either metabolic wastes or substances with no role in fundamental life processes; they were also thought to be unusual compounds limited in distribution among plants. Although such sweeping definitions are no longer used, the label remains and adds to the confusion associated with secondary products.

The evolution of phenolic compounds may have been in response to the lack of motility in plants. These compounds may have offered protection from predators and from ultraviolet (UV) radiation, which was much more intense when higher plants were evolving.[1] Phenolics in particular absorb far (40 to 280 nm) and midrange (280 to 320 nm) UV, which are damaging wavelengths to proteins and nucleic acids. They also act as radical quenchers of stress-created singlet oxygen.[2] It seems likely that any new secondary product that did not confer some kind of advantage to the plant would eventually have been lost through natural selection, unless it was closely associated on a chromosome with a highly favorable characteristic. Therefore, most of the so-called secondary products found in present day plants probably have functions associated with increasing survival, either by coping with unfavorable environmental conditions, including predation, or by regulating metabolic processes.

Loosely speaking, phenolic compounds are those containing a benzene ring with one or more hydroxyl groups; more specifically, they are derivatives of the shikimic acid pathway which also produces phenylalanine and tyrosine. Phenolics are ubiquitous in higher plants and occur in some lower plants as well. Usually phenolics are esterified with sugar moieties or are likewise detoxified in the intact plant; most of them are bound to the cell wall or are stored in the vacuole. However, disruption of the cell by environmental stress or laboratory tissue homogenization releases these compounds, which then dissociate enzymatically from their ester linkages and will quickly modify cell components such as enzymes and proteins.

Chemical modification by phenolics of proteins may occur by any of four mechanisms as shown in Figure 1:

1. **H-bonding** — The phenolic hydroxyl group(s) can hydrogen bond to the δ^- carbonyl oxygen of peptide bonds. This is a very strong bond that cannot be dissociated by dialysis or gel filtration.[3]
2. **Covalent coupling** — Phenolics can quickly oxidize to form quinones enzymatically or under high pH conditions (pH 7 and up). Quinones are powerful oxidants which readily polymerize and condense with sulfhydryl, amino, or imino groups.[4,5]
3. **Ionic interactions** — At high pHs (pH 8 and up), phenolics can also form salt bridges with basic amino acid residues.[5,6]
4. **Hydrophobic interactions** — The aromatic ring structure of phenolics will interact with hydrophobic protein regions.[5,6]

By these given mechanisms, phenolics and their oxidation products can effectively tie up proteins, modify enzymatic active sites and thus decrease activity, and even create artifactual enzymes by blocking ionic amino acid residues such as lysine.

II. PHENOLIC BIOSYNTHESIS

A. Enzymatic Regulation

The first committed step in phenolic biosynthesis is the removal of ammonia from L-phenylalanine by phenylalanine amonia-lyase (PAL). The product, *trans*-cinnamic acid, is the precursor for phenolic components of lignin, suberin, tannins, flavonoids, coumarins, and other compounds as illustrated in Figure 2. The regulation of PAL, then, is of critical

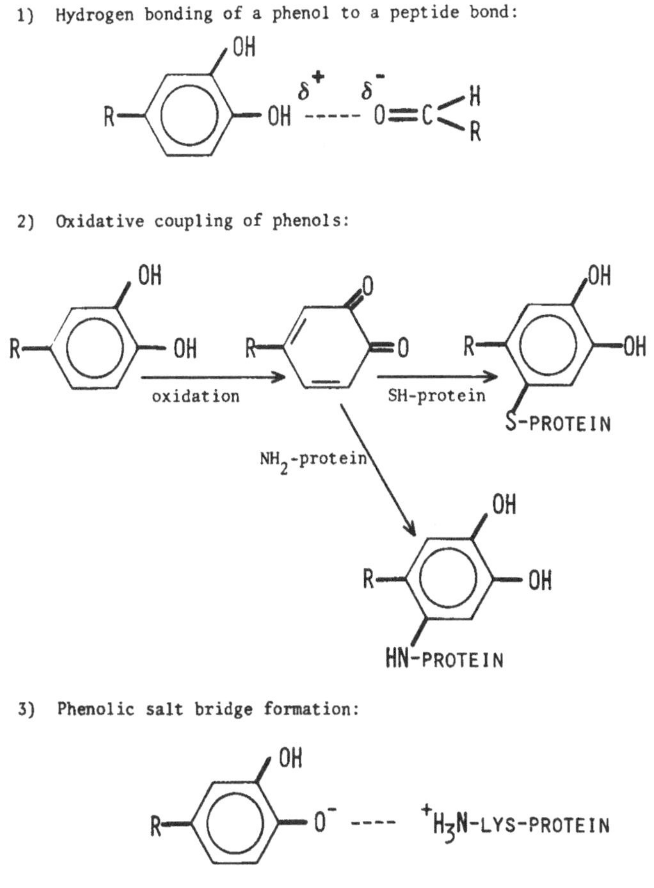

FIGURE 1. Chemical modification by reactive phenolic compounds.

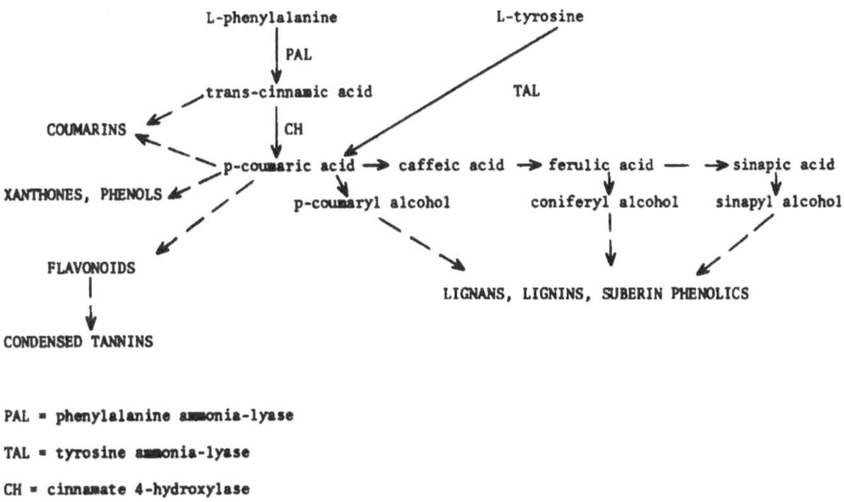

FIGURE 2. An outline of the biosynthesis of some selected plant phenolic compounds.

importance in any metabolic activity in the plant which requires the synthesis of such compounds. (In monocots and a few dicots, tyrosine ammonia-lyase (TAL) deammonifies L-tyrosine to form *p*-coumaric acid. In most higher plants, however, TAL is of secondary importance and may not exist at all.)

PAL is a phytochrome-controlled, tetrameric enzyme of mol wt 333,000. It may occur as several isozymes, each inhibited by a different phenolic, such as cinnamic or *p*-coumaric acid. Reportedly it exists in the lumen of endoplasmic reticulum membranes in the cytosol and in the stroma of chloroplasts,[7] and is subject to rapid turnover in vivo.[8]

The regulation of PAL *de novo* synthesis and activity has been the subject of much research. The effect of visible light upon PAL is generally to increase synthesis and activity;[7] similarly, UV-irradiation was seen to increase mRNA for PAL synthesis in parsley culture.[9] Other stresses, such as wounding,[10,11] thigmomorphogenesis,[12] ethylene treatment,[13] nutrient deficiencies,[14] herbicide treatment,[15] and viral,[7] fungal,[11,16] or insect attack[17] have also been shown to increase either PAL synthesis or activity in a variety of plants. In contrast, stress events do not usually inhibit PAL synthesis or activity. Inhibition occurs either by competitive substrate inhibitors of L-phenylalanine (including D-phenylalanine and 2-aminooxy-3-phenylpropionic acid [AOPP]) and by feedback inhibition of product derivatives (including cinnamic acid and *p*-coumaric acid).

The next enzyme in phenolic biosynthesis is cinnamate-4-hydroxylase. It is an irreversible cytochrome P_{450} oxygenase located on the face of the chloroplast and ER membranes. The other well-characterized hydroxylase is ferulic acid 5-hydroxylase. It too, is a cytochrome P_{450} oxygenase and is thought to be located in the microsomal fraction.[18] Methyl transferases, utilizing s-adenosyl methionine (SAM), are required for methylation of the phenolic acids, which then may become their respective alcohols via 4-coumarate:CoA ligase.

The enzyme responsible for incorporating phenolic precursors into lignin and suberin is peroxidase. It is located within the cell walls, where it attaches the phenolic alcohols by oxidizing NADH and forming peroxide.[19] Not suprisingly, peroxidase activity is increased by thigmomorphogenesis and ethylene.[12] Polyphenol oxidase, which will link groups of phenolic acids, is also stimulated by ethylene[20] and fungal infection.[11]

Very little is known of the biodegradation of phenolic compounds. Fungal enzymes are known to degrade phenolics, but mediated turnover within the intact plant is not well documented. Floral and leaf flavonoid pigments are known to turn over quickly, but relatively little else has been discovered about the nature of autodegradation of other phenolic polymers.

B. Cell-Wall Phenolics

One of the products of phenolic biosynthesis is lignin, a cell-wall component of all vascular plants. Lignin precursors are formed in the endoplasmic reticulum, packaged into vesicles, and transported to the cell walls.[21] Polymerization occurs within the wall and requires the presence of free radicals via the peroxidase/ascorbate system. Although considered to be a normal cell-wall thickening process, lignification can also be induced by UV-irradiation,[22] herbicide application,[23] and thigmomorphogenesis.[12] Wounding stress,[11,24,25] nutrient stress,[26] fungal infection,[11,27,28] and ethylene[12,29,30] had lignifying effects on several plant species. Cold temperatures can also induce lignification; this relationship will be discussed in detail later.

A second defensive barrier within the cell wall is suberin. Although its exact structure and composition is not known, as pure suberin is difficult to isolate, it has both phenolic and polyester components. In contrast to the lignin phenolics, suberin phenolics generally have fewer substitutions and fewer *o*-methoxy groups.[31] The polyester component of suberin appears to consist of ω-hydroxy fatty acids and dicarboxylic acids.[32] Originally thought to occur only in plant roots, suberin serves to reduce moisture loss through the cell wall. It also acts as a barrier to solutes and microbes, and can be formed as a general response to stress.

Wound-induced suberin or its precursors have been found in jade leaves, tomatoes, and beans;[33] potato tubers;[34] peach tree bark,[24,35,36] and twigs.[37] Similarly, suberin induction occurs in response to fungal elicitors in carrot slices[38] and cold temperatures in white spruce roots.[39] Cold temperatures generally induce cell-wall thickening; Griffith[40,41] noticed an increase in cell-wall thickness toward the vascular bundles and leaf surfaces in cold acclimated winter rye. Associated with this was an increase in the production of lipids, hypothesized to be of suberin character. Paroschy et al.[42] found the lack of adequate suberization to be the cause of frost damage to late-maturing grape varieties. Further discussion of suberization and its role in cold acclimation will occur later in this chapter.

III. FUNCTIONS OF PHENOLIC COMPOUNDS

A. Growth Regulation

Phenolic compounds are also involved in the growth status of plants. Seeds of some plants are protected from premature germination by phenolic growth inhibitors in their outer layers,[43-45] thus reducing the possibility of germination during environmentally unfavorable conditions. The mechanism by which germination is inhibited often involves the enzymatic formation of tannins and lignins within the coat,[46,47] which would not only make the coat impervious to water, but would also react with any available atmospheric oxygen. The chemical inhibition of polyphenol oxidase was shown to overcome the dormancy of barnyard grass seed.[48]

Generally, seed coat phenolics decrease during stratification,[43,49,50] usually with no effect on seed germination. It might be concluded, therefore, that their function is not continuously associated with maintaining seed dormancy. Instead, because they are in the outer layer of the seed, it is possible that they initially prevent germination by restricting gas and water exchange through the seed, and also protect the embryo from environmental stresses and disease. For instance, fungally infected strawerry seeds were seen to have accelerated germination, possibly due to the removal of phenolic inhibitors.[51]

In contrast, the presence of phenolics in vegetative parts of the plant may actively inhibit or stimulate growth and development. Depending on their potential reactivity with proteins and enzymes, phenolic are classified as either "growth inhibitory" or "growth stimulatory." The stimulatory *o*-dihydroxy and trihydroxyphenols such as caffeic acid have highly reactive hydroxyl groups in a conformation that can inactivate IAA-oxidase, thus inhibiting IAA decarboxylation. Conversely, the monohydroxyphenolics inducing *p*-coumaric acid and the nonhydroxylated cinnamic acids do not have the obligatory adjacent hydroxyl groups, and are therefore not inimical to IAA-oxidase activity; hence, they are considered to be inhibitory to growth. There is considerable confusion over these designations in the literature, primarily because phenolics which can regulate IAA-oxidase activity can also presumably regulate other enzymes as well, which may or may not inhibit growth.

Studies conducted on buds or other vegetative plant parts generally support the idea of phenolic participation in plant growth.[52] Endogenous phenolics were found to be major determinants of IAA-oxidase activity in douglas fir seedlings,[53] tomato callus,[54] and lentils.[55] Lavee et al.[56] suggest that chlorogenic acid (a growth stimulatory phenolic) can compensate for IAA in olive callus cultures. Other mechanisms of phenolic influence in plant growth have been postulated. Ray and Laloraya[57] found that phenolics were antagonistic to ABA action and would restore seedling growth in *Amaranthus*. In contrast, Rodriguez and Sanchez-Tames[58] found phenolics to enhance the effect of ABA in maintaining the dormancy of hazel buds.

B. Disease Resistance

The interaction between plants and biotic elicitors (fungi, bacteria, viruses, nematodes) or abiotic elicitors (UV-irradiation, freezing, heating, wounding, chemicals) often results in the production of phytoalexins.[59,60] Although phytoalexins were originally defined to be antimicrobial secondary metabolites which are synthesized in respones to microbial infection, their elicitation by other factors demands that the definition be broadened. Furthermore, phytoalexin production is not the only method by with plants may resist disease. As there have been numerous reviews on phytoalexins and their role in plant defense,[60-62] this discussion will focus primarily upon other mechanisms of disease resistance.

The general nature of phytoalexin induction is suggested by the appearance of "reaction zones" in tree sapwood subjected to wounding injury or fungal attack.[63] Similar barriers of disease-resistant, phenolic impregnated cell walls have been found in bean,[64] soybean,[65] cucumber,[66] tobacco,[28] wheat,[27] beech,[67] banana root,[68] and peach trees[35,69,70] as a result of wounding and/or fungal infection.

In some cases, there appears to be "immunization" against disease by a previous infection or wounding;[68,71] this concept is illustrated in the work by Ostrofsky et al.[67] Wound altered bark of fungus-resistant beech trees was found to have a higher phenolic content than that of susceptible beech; the healthy bark of both trees were equal in their phenolic content. Similarly, peach trees that had been wounded and then subjected to fungal infection showed an increase in phenolics over and above that due to the wounding response.[70] Herbicide exposure of potatoes increased lignification while reducing their susceptibility to fungal infection.[23]

Naturally resistant plants may have inherently higher phenolic contents; for instance, the roots of rot-resistant species including pecan, persimmon, and passionfruit do have higher phenolic levels than susceptible species such as apple, almond, and peach.[72] Such common phenolic compounds as ferulic, *o*-coumaric, and *p*-coumaric acids inhibit fungal growth in vitro;[73] onion bulbs containing catechol and protocatechuic acid are naturally fungal resistant.[74] The basis for resistance may be the activity of enzymes responsible for phenolic synthesis. PAL, polyphenol oxidase, and peroxidase activities have been implicated in disease resistance.[62,75] Additionally, treatments that reduce or inhibit PAL synthesis or activity will reduce resistance to pathogens.[60]

C. General Stress Responses

In addition to acting as protective agents against fungal stress, phenolic concentrations are affected by other types of stresses as well. In general, any kind of mechanical wounding will lead to increased phenolic synthesis[11,12,76] and accumulation,[77-79] especially in the cell walls.[12,24,25,80]

Nutrient stress also has a marked effect on phenolic levels in plant tissues. Deficiencies in nitrogen and phosphate lead to accumulation of phenolics[81,82] and lignification.[26] Shortages of potassium and sulfur are reported to increase phenolic concentrations;[14] water deficit has variable effects,[14] but has been seen to increase root suberization.[83] Magnesium deficiency is also reported to increase suberization,[32] while iron deficiency inhibits the process.[84] Interestingly enough, high concentrations of glucose[85] and sucrose[86] will also stimulate phenolic production.

Other stresses found to increase phenolic accumulation include cadmium toxicity,[87] UV irradiation,[22] and heat shock.[88] The effects of cold stress upon plant phenolic levels is considered later.

It is possible that phenolic synthesis and accumulation is a generalized stress response, perhaps occurring at the membrane level. Walker-Simmons et al.[89] found that wounded potato and tomato leaves developed weakened cell membranes as a response to stress. Fungal elicitors were seen to depolarize membranes by production of ethylene;[90] similarly, Glass and Dunlop[91] found membranes were depolarized by the addition of phenolics to cell cultures. In both cases, membranes were made more permeable to ions. Herbicide treatment of crops also increased membrane permeability, while simultaneously increasing PAL activity and generating ethylene.[15] The release of either ethylene or phenolics could act as a general elicitor to the cell's defense mechanisms.

D. Interactions With Ethylene

There exists in the literature a great deal of evidence that ethylene production and phenolic synthesis are closely related. To determine which compound elicits the synthesis of the other, however, is difficult to answer. Mapson[92] thought *p*-coumaric acid to be a necessary co-factor in the synthesis of ethylene during post-harvest fruit development; later Wardale[93] expanded upon this hypothesis by adding naringenin as a ethylene inducer. He hypothesized that these two phenolic compounds are substrates for peroxidase activity leading to ethylene production, and that other endogenous phenolics were inhibitory. Similarly, Yang[94] found that peroxidase catalysis of monohydroxy phenolics (such as *p*-coumaric acid) did lead to the formation of ethylene, but dihydroxy phenolics were inhibitory to its biosynthesis. An unknown phenolic is reported to inhibit the synthesis of ethylene from ACC in carnation petals as well.[95]

More recent information confirming these hypotheses is provided by Billot's[96] research on the post-harvest physiology of pears. He observed that the ratio of monohydroxy to dihydroxy phenolics reached a maximum as ethylene synthesis increased during ripening. As mentioned previously, monohydroxy phenolics are considered to be IAA-oxidase sparing; therefore this ripening period would also have maximal IAA-oxidase activity. This observation correlates well with the results of Vioque et al.,[97] who postulated that the splitting of the ACC ring structure to form ethylene is dependent upon IAA-oxidase-generated hydrogen peroxide.

There is convincing evidence, however, that ethylene can induce the production of phenolic compounds as well. Ethylene has been seen to increase PAL activity and/or synthesis in carrot,[13] lettuce,[98] and in various fruit and vegetable tissue.[30] It also increases the activity of other enzymes in the phenolic pathway including peroxidase,[12] polyphenol oxidase,[20] and cinnamate 4-hydroxylase:*p*-coumarate CoA ligase.[30] Lignification has been reported to increase as a result of ethylene stimulation.[12,29,30]

Stress conditions, such as the presence of cadmium, can induce the synthesis of ethylene and activate phenolic enzymes.[87] In wounding studies performed on cottonwood[79] and poplar leaves,[77] increases in phenolics were seen not only in the wounded trees, but in undamaged neighboring trees as well. The presence of an airborne, transmitted signal is suggested; this signal could conceivably be ethylene.

E. Cold Stress Response

One of the simplest ways of assessing cold injury in plant tissues is by observing the typical browning associated with membrane rupture. As compartmentalized phenolic compounds and enzymes are simultaneously released into the cell, the phenolics lose their esterified moieties and are then free to react with proteins, forming insoluble brown complexes. Under acidified conditions ($pH < 5.5$), these oxidative reactions are somewhat inhibited. Spectrophotometric measurement[99] of water-soluble phenolics leached from freeze-injured tissues correlates well with other viability methods in determining plant injury (Table 1).

In rhododendron leaf tissues that have been sublethally cold stressed, a net accumulation of water-soluble phenolic compounds was found (unpublished data). Analysis of these phenolics, extracted with acetone and subjected to two-dimensional thin layer chromotography (TLC), reveals the appearance of new compounds that are not found in either nonfrozen controls or lethally freeze-stressed tissues. Because of their staining properties and mobility in the TLC solvents, they are tentatively identified as being small phenolic acids esterified with a sugar moiety, pending further analysis.

Table 1
CORRELATION COEFFICIENTS BETWEEN % PHENOLIC LEAKAGE AND OTHER METHODS OF FREEZE-DAMAGE EVALUATION AT DIFFERENT TIMES OF THE YEAR.[99]

	Method			
Date	% Cond.	Browning	Ethane	TTC Red.
September 1985	0.99[b]	0.99[b]	0.95[a]	−0.88[a]
January 1986	0.99[b]	0.99[b]	0.96[a]	—
April 1986	0.99[b]	0.99[b]	0.99[b]	—

Note: Significance determined by using t-values where:

$$t = \sqrt{\frac{rn - 2}{1 - r^2}}$$

[a] Significant at 5%.
[b] Significant at 1%.

Following a similar freeze-stress experiment, radio-labeled ^{14}C-phenylalanine was added to rhododendron leaf discs. Preferential incorporation of the label was seen in the cell-wall fraction of sublethally stressed tissues compared to nonfrozen controls. Additionally, there was an increase in the labeled protein fraction (which includes enzymes) in these same stressed tissues (unpublished data).

We hypothesize that cold stress increases phenolic production and subsequent incorporation into the cell wall either as suberin or lignin. Although futher experimentation is needed to validate these results, evidence in the literature supports the concept of phenolic involvement in cold stress response and acclimation. The synthesis of PAL is seen to increase following chilling of apple fruit and potatoes,[100] leading to an increase in phenolics. In addition, apple tree adaptation to cold climates was found to be associated with high levels of chlorogenic acid.[101]

It has been concluded that lignification serves to increase the frost tolerance of grapevine,[102,103] apple trees,[104] and sugarcane.[105] Similary, suberin deposition is also shown to increase resistance to cold temperatures. Suberization of white spruce roots occurred following cold temperature regimes.[39] During cold acclimation of winter rye, Griffith et al.[41] noticed thickened cell walls and increased production of suberin-type lipids. Frost-hardened pine seedlings were also seen to increase their lipid synthesis.[106] The lack of cold hardiness in late maturing grape was also attributed to inadequate suberization.[42]

A mechanism by which suberin and lignin may protect plants from freeze damage has been suggested by Griffith et al.[41] and Paroschy et al.:[42] the deposition of cell wall materials could maintain the water status of tissues by acting as a water barrier. As plants acclimate, their water content decreases; if they have no mechanism by which to keep their freezable water below critical levels during thaws, they can be killed by subsequent frosts. In addition, effective water barriers would help reduce evapotranspiration resulting from extracellular freezing, and thus decrease injury due to desiccation. This may account for the increased thickening seen in cell walls next to vascular bundles and towards leaf surfaces in cold acclimated winter rye.[40]

The importance of cell-wall associated phenolic substances in the process of cold acclimation is further enhanced by research by Bartolo and Wallner.[107] They noted that cold acclimation increases the resistance to cellulase in leaf tissues, as does wounding; wounding also induces cold hardiness in the tissues. A possible change at the cell-wall membrane

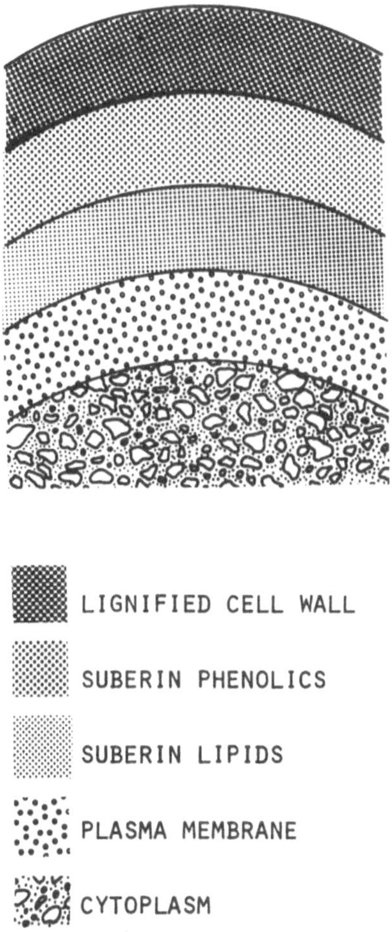

FIGURE 3. Stylized cross-section depicting cell wall layers of a hypothetical freeze-resistant plant.

interface resulting in increased membrane adhesion to the cell wall could reduce cell-wall collapse during freeze-induced dehydration.

In Figure 3, a stylized cell wall is shown. It is quite likely that the outer layers of the cell are not as distinct as once was assumed. The primary cell wall, consisting of cellulose and other sugars, has been recently found to contain cinnamate,[108] ferulic acid, and *p*-coumaric acid[109] which are normally associated with the secondary cell wall. They could serve as insoluble linkages between cellulose and secondary cell-wall materials. Similarly, the location of suberin in the cell wall provides phenolic attachments toward the outer layers, and lipid attachments towards the membranes. This orientation could well account for the speculated increased membrane-cell wall adhesion suggested by Bartolo and Wallner.[107]

IV. CONCLUSIONS

Phenolic compounds are involved in plant responses to environmental stresses including wounding, pathogen attack, mineral deficiencies, and temperature stress. The response of resisting plants is apparently to increase either phenolic synthesizing enzyme levels or their activities, leading to increased phenolic production. These reactive compounds may then modify the developmental status of the plant, either independently or by interaction with known plant growth substances, especially ethylene. Additionally, these compounds are often the precursors for lignin and suberin, and may be polymerized into the cell wall. Such cell-wall thickenings are not only more resistant to pathogen attack, they may also protect the plant from freezing stress. An increase in points of attachment between the plasma membrane and the rigid cell wall could reduce cell collapse during freezing-induced dehydration and mechanical stress, and thus increase the freezing resistance of the plant.

REFERENCES

1. **Lowry, B., Lee, D., and Hébant, C.,** The origin of land plants: a new look at an old problem, *Taxon,* 29, 183, 1980.
2. **Knox, J. P. and Dodge, A. D.,** Singlet oxygen and plants, *Phytochemistry,* 24, 889, 1985.
3. **Loomis, W. D. and Battaile, J.,** Plant phenolic compounds and the isolation of plant enzymes, *Phytochemistry,* 5, 423, 1966.
4. **Laties, G. G.,** Isolation of mitochondria from plant material, in *Methods in Enzymology, Vol. 31, Biomembranes, Part A,* Fleischer, S. and Parker, L., Eds., Academic Press, New York, 1974, 589.
5. **Loomis, W. D.,** Overcoming problems of phenolics and quinones in the isolation of plant enzymes and organelles, in *Methods in Enzymology, Vol. 31, Biomembranes, Part A,* Fleischer, S., and Parker, L., Eds., Academic Press, New York, 1974, 528.
6. **Loomis, W. D.,** personal communication, 1984.
7. **Jones, D. H.,** Phenylalanine ammonia-lyase: regulation of its induction and its role in plant development, *Phytochemistry,* 23, 1349, 1984.
8. **Davies, M. E.,** Effects of auxin on polyphenol accumulation and the development of phenylalanine ammonia-lyase activity in darkgrown suspension cultures of Paul's scarlet rose, *Planta,* 104, 66, 1972.
9. **Schröder, J., Kreuzaler, F., Schäfer, E., and Hahlbrock, K.,** Concomitant induction of phenylalanine ammonia-lyase and flavanone synthesis mRNAs in irradiated plant cells, *J. Biol. Chem.,* 254, 57, 1979.
10. **Pascholati, S. F., Nicholson, R. L., and Butler, L. G.,** Phenylalanine ammonia-lyase activity and anthocyanin accumulation in wounded maize mesocotyls, *J. Phytopathol.,* 115, 165, 1986.
11. **Thorpe, J. R. and Hall, J. L.,** Chronology and elicitation of changes in peroxidase and phenylalanine ammonia-lyase activities in wounded wheat leaves in response to inoculation by *Botrytis cinerea, Physiol. Plant Pathol.,* 25, 36, 1984.
12. **de Jaegher, G., Boyer, N., and Gaspar, T.,** Thigmomorphogenesis in *Bryonia dioica:* changes in soluble and wall peroxidases, phenylalanine ammonia-lyase activity, cellulose, lignin content and monomeric constituents, *Plant Growth Reg.,* 3, 133, 1985.
13. **Chalutz, E.,** Ethylene-induced phenylalanine ammonia-lyase activity in carrot roots, *Plant Physiol.,* 51, 1033, 1973.
14. **Gerschenzon, J.,** Changes in the levels of plant secondary metabolites under water and nutrient stress, in *Phytochemical Adaptations to Stress,* Timmermann, B. N., Steelink, C., and Loewus, F. A., Eds., Plenum Press, New York, 1983, 273.
15. **Kömives, T. and Casida, J. E.,** Acifluorfen increases the leaf content of phytoalexins and stress metabolites in several crops, *J. Agric. Food Chem.* 31, 751, 1983.
16. **Bolwell, G. P., Cramer, C. L., Lamb, C. J., Schuch, W., and Dixon, R. A.,** L-Phenylalanine ammonia lyase from *Phaseolus vulgaris:* modulation of the levels of active enzyme by trans-cinnamic acid, *Planta,* 169, 97, 1986.
17. **Cole, R. A.,** Phenolic acids associated with the resistance of lettuce cultivars to the lettuce root aphid, *Ann. Appl. Biol.,* 105, 129, 1984.
18. **Grand, C.,** Ferulic acid 5-hydroxylase: a new cytochrome P-450-dependent enzyme from higher plant microsomes involved in lignin synthesis, *FEBS Lett.,* 169, 7, 1984.
19. **Mäder, M. and Füssl, R.,** Role of peroxidase in lignification of tobacco cells. II. Regulation by phenolic compounds, *Plant Physiol.,* 70, 1132, 1982.
20. **van Lelyveld, L. J. and Nel, E.,** Ethylene concentration and polyphenol oxidase activity in mango *(Mangifera indica* L.) fruit abscission, *Z. Pflanzen Physiol.* 107, 179, 1982.
21. **Northcote, D. H.,** Cell wall biosynthesis and assembly, Proc. Phytochem. Soc. Europe, Bangor, Wales, 1985.
22. **Zhakote, A. G., Negru, P. V., and Gaugash, M. V.,** The effect of the duration and spectral composition of natural light on grapevine photosynthesis and frost resistance, *Fiziol. Biokhim.,* 55, 1979.

23. **Barak, E., Edgington, L. V., and Ripley, B. D.,** Bioactivity of the fungicide metalaxyl in potato tubers against some species of *Phytophthora, Fusarium,* and *Alternaria,* related to polyphenol-oxidase activity, *Can. J. Plant Pathol.,* 6, 304, 1984.
24. **Biggs, A. R.,** Phellogen regeneration in injured peach tree bark, *Ann. Bot.,* 57, 463, 1986.
25. **Rickard, J. E. and Gahan, P. B.,** The development of occlusions in cassava (*Manihot esculenta* Crantz) root xylem vessels, *Ann. Bot.,* 52, 811, 1983.
26. **Farmer, E. E.,** Effects of fungal elicitor on lignin biosynthesis in cell suspension cultures of soybean *Plant Physiol.,* 78, 338, 1985.
27. **Beardmore, J., Ride, J. P., and Granger, J. W.,** Cellular lignification as a factor in the hyper-sensitive resistance of wheat to stem rust, *Physiol. Plant Pathol.,* 22, 209, 1983.
28. **Zhou, J. P. and Wang, B. C.,** Field observation on the epiphytology of tobacco black shank *(Phytophthora parasitica* f. *nicotianae* [Breda de Haan] Tucker), *Acta Phytopathol. Acad. Sci. Hung.,* 14, 39, 1984.
29. **Fouad, M. K., Nagdy, G. A., and Mahmoud, W. S.,** Effect of Cyclocel and Ethrel treatments on the anatomical structure of stems and leaves of pepper plant, *Capsicum annuum* L., *Ain Shams Univ. Fac. Agric. Bull.,* 1153, 1979.
30. **Rhodes, M. J. C. and Wooltorton, L. S. C.,** Phenolic metabolism in fruit and vegetable tissue under stress, *ARC Food Research Institute Biennial Report 1977-1978,* Norwich, UK, 1980.
31. **Kolattukudy, P. E.,** Structure, biosynthesis and biodegradation of cutin and suberin, *Annu. Rev. Plant. Physiol.,* 32, 539, 1981.
32. **Kolattukudy, P. E.,** Biochemistry and function of cutin and suberin, *Can. J. Bot.,* 62, 2918, 1984.
33. **Dean, B. B. and Kolattukudy, P. E.,** Synthesis of suberin during wound-healing in jade leaves, tomato fruit, and bean pods, *Plant Physiol.,* 58, 411, 1976.
34. **Cottle, W. and Kolattukudy, P. E.,** Biosynthesis, depostion, and partial characterization of potato suberin phenolics, *Plant Physiol.,* 69, 393, 1982.
35. **Biggs, A. R.,** Boundary-zone formation in peach bark in response to wounds and *Cytospora leucostoma* infection. *Can. J. Bot.,* 62, 2814, 1984.
36. **Biggs, A. R. and Miles, N. W.,** Suberin deposition as a measure of wound response in peach bark, *HortScience,* 20, 903, 1985.
37. **Biggs, A. R. and Northover, J.,** Formation of the primary protective layer and phellogen after leaf abscission in peach, *Can. J. Bot.,* 63, 1547, 1985.
38. **Heale, J. B., Dodd, K. S., and Gahan, P. B.,** The induced resistance response of carrot root slices to heat-killed conidia and cell-free germination fluid of *Botrytis cinerea* Pers. ex. Pers. I. The possible role of cell death, *Ann. Bot.,* 49, 847, 1982.
39. **Johnson-Flanagan, A. M. and Owens, J. N.,** Peroxidase activity in relation to suberization and respiration in white spruce *(Picea glauca* [Moench] Voss) seedling roots, *Plant Physiol.,* 79, 103, 1985.
40. **Griffith, M. and Brown, G. N.,** Cell wall deposits in winter rye *Secale cereale* L. 'Puma' during cold acclimation, *Bot. Gaz. (Chicago),* 143, 486, 1982.
41. **Griffith, M., Huner, N. P. A., Espelie, K. E., and Kolattukudy, P. E.,** Lipid polymers accumulate in the epidermis and mestome sheath cell walls during low temperature development of winter rye leaves, *Protoplasma,* 125, 53, 1985.
42. **Paroschy, J. H., Meiering, A. G., Peterson, R. L., Hostetter, G., and Neff, A.,** Mechanical winter injury in grapevine trunks, *Am. J. Enol. Viti.,* 31, 227, 1980.
43. **Fadl-Mostafa, S., Baz-Abdel-Ghany, I. O., and Tayel, S.,** Dormancy termination in apricot seeds. II. Interaction between endogenous phenols and germination, *Egypt. J. Physiol. Sci.,* 5, 139, 1980.
44. **Cezard, R.,** Orobanchaceae. II. Breaking dormancy in orobanchanceous seeds by removing an endogenous inhibition, *Bull. Acad. Soc. Lorraine Sci.,* 12, 97, 1973.
45. **Naqvi, H. H. and Hanson, G. P.,** Germination and growth inhibitors in guayule *(Parthenium argentatum* Gray) chaff and their possible influences in seed dormancy, *Am. J. Bot.,* 69, 985, 1982.
46. **Egley, G. H., Paul, R. N., Jr., Duke, S. O., and Vaughn, K. C.,** Peroxidase involvement in lignification in water-impermeable seed coats of weedy leguminous and malvaceous species, *Plant Cell Environ.,* 8, 253, 1985.
47. **Aparicio-Tejo, P. M., Peña, J. I., Sánchez-Diaz, M., and Becana, M.,** Coat phenolic compounds inhibit germination of *Halimium halimifolium* L., *Plant Physiol.,* Suppl. 80 (Abstr.), 128, 1986.
48. **Shimizu, N. and Ueki, K.,** Studies on the breaking of dormancy in barnyard grass seed. III. Change of the dormancy-breaking effect of various compounds concerned with oxidation-reduction system by dormancy stage, *Proc. Crop Sci. Soc. Japan,* 41, 488, 1972.
49. **Chen, F. S., MacTaggart, J. M., and Elofson, R. M.,** Chemical consitituents in wild oat *(Avena fatua)* hulls and their effects of seed germination, *Can. J. Plant Sci.,* 62, 155, 1982.
50. **Murphy, J. B. and Noland, T. L.,** Changes in phenolic acids and abscisic acid in sugar pine seed coats during stratification, *Physiol. Plant.,* 52, 370, 1981.
51. **Guttridge, C. G., Woodley, S. E., and Hunter, T.,** Accelerating strawberry seed germination by fungal infection, *Ann. Bot.,* 54, 223, 1984.

52. **Kefeli, V. I. and Kutacek, M.**, Phenolic substances and their possible role in plant growth regulation, in *Plant Growth Regulation*, Pilet, P. E., Ed., Springer-Verlag, New York, 1977, 181.
53. **Johnson, M. A. and Carlson, J. A.**, Indoleacetic acid oxidase and related enzymes in cultured and seedling Douglas fir., *Bicheem., Physiol. Pflanz.*, 174, 115, 1979.
54. **Ranade, S. and David, S. B.**, Quinones as plant growth regulators, *Plant Growth Reg.*, 3, 3, 1985.
55. **Dangar, T. K. and Basu, P. S.**, Seasonal changes and metabolism of plant hormones in root nodules of *Lens* sp., *Biol. Plant.*, 26, 253, 1984.
56. **Lavee, S., Avidan, N., and Harshemesh, H.**, Chlorogenic acid, an alternative or reverse pathway for IAA in the olive, Int. Congr. Plant Tissue Cell Culture Annu. Meeting, 1985.
57. **Ray, S. D. and Laloraya, M. M.**, Interaction of gibberellic acid, abscisic acid, and phenolic compounds in the control of hypocotyl growth of *Amaranthus caudatus* seedlings, *Can. J. Bot.*, 62, 2047, 1984.
58. **Rodriguez, A. and Sánchez-Tamés, R.**, Dormancy and seasonal changes of plant growth regulators in hazel buds, *Physiol. Plant.*, 66, 288, 1986.
59. **Bailey, J. A.**, Mechanisms of phytoalexin accumulation, in *Phytoalexins*, Bailey, J. A. and Mansfield, J. W., Eds., John Wiley & Sons, New York, 1982, 289.
60. **Darvil, A. G., and Albersheim, P.**, Phytoalexins and their elicitors — a defense against microbial infection in plants, *Annu. Rev. Plant Physiol.*, 35, 243, 1984.
61. **Bailey, J. A. and Mansfield, J. W., Eds.**, *Phytoalexins*, John Wiley & Sons, New York, 1982.
62. **Friend, J.**, Plant phenolics, lignification, and plant disease, in *Progress in Phytochemistry*, Reinhold, L., Harborne, J. B., and Swain, T., Eds., Pergamon Press, New York, 1981, 197.
63. **Kemp, M. S. and Burden, R. S.**, Phytoalexins and stress metabolites in the sapwood of trees, *Phytochemistry*, 25, 1261, 1986.
64. **Stumpf, M. A. and Heath, M. C.**, Cytological studies of the interactions between the cowpea rust fungus and silicon-depleted French bean plants, *Physiol. Plant Pathol.*, 27, 369, 1985.
65. **Köhle, J., Young, D. H., and Kauss, H.**, Physiological changes in suspension-cultured soybean cells elicited by treatment with chitosan, *Plant Sci. Lett.*, 33, 221, 1984.
66. **Hammerschmidt, R., Bonnen, A. M., Bergstrom, G. C., and Baker, K. K.**, Association of epidermal lignification with nonhost resistance of cucurbits to fungi, *Can. J. Bot.*, 63, 2393, 1985.
67. **Ostrofsky, W. D., Shortle, W. C., and Blanchard, R. O.**, Bark phenolics of American beech *(Fagus grandifolia)* in relation to the beech bark disease, *Eur. J. For. Pathol.*, 14, 52, 1984.
68. **Beckman, C. H., Mueller, W. C., and Mace, M. E.**, The stabilization of artificial and natural cell wall membranes by phenolic infusion and its relation to wilt disease resistance, *Phytopathology*, 64, 1214, 1974.
69. **Biggs, A. R.**, Wound age and infection of peach bark by *Cytospora leucostoma*, *Can. J. Bot.*, 64, 2319, 1986.
70. **Wisniewski, M., Bogle, A. L., Shortle, W. C., and Wilson, C. L.**, Interaction between *Cytospora leucostoma* and host-phenolic compounds in dormant peach trees, *J. Am. Soc. Hort. Sci.*, 109, 563, 1984.
71. **Kuc, J.**, Induced immunity to plant disease, *BioScience*, 32, 854, 1982.
72. **Sztejnberg, A., Azaizia, H., and Chet, I.**, The possible role of phenolic compounds in resistance of horticultural crops to *Dematophora nexatrix* Hartig, *Phytopathol. Z.*, 107, 318, 1983.
73. **Perradin, Y., Mottet, M. J., and Lalonde, M.**, Influence of phenolics of *in vitro* growth of *Frankia* strains, *Can. J. Bot.*, 61, 2807, 1983.
74. **Deverall, B. J.**, Introduction, in *Phytoalexins*, Bailey, J. A. and Mansfield, J. W., Eds., John Wiley & Sons, New York, 1982, 1.
75. **Bell, A. A.**, Biochemical mechanisms of disease resistance, *Annu. Rev. Plant Physiol.*, 32, 21, 1981.
76. **van Loon, L. C. and Gerritsen, Y. A. M.**, Ethylene and a product of aromatic biosynthesis induce pathogenesis-related proteins in hypersensitively reacting tobacco, *Plant Physiol.*, Supp. 80 (Abstr.), 145, 1986.
77. **Baldwin, I. T. and Schultz, J. C.**, Rapid changes in tree leaf chemistry induced by damage: evidence for communication between plants, *Science*, 221, 277, 1983.
78. **Barker, W. G. and Peterson, R. L.**, Wound responses of *Impatiens balsamina* seedlings after collet excision, *Scanning Electron Microsc.*, 1001, 1984.
79. **Kimmerer, T. W.**, Induction of phenol glycoside synthesis in eastern cottonwood, *Phytochemistry*, in press, 1988.
80. **Biddington, N. L.**, A review of mechanically induced stress in plants, *Sci. Hortic.*, 36, 12, 1985.
81. **DiCosmo, F. and Towers, G. H. N.**, Stress and secondary metabolism in cultured plant cells, in *Phytochemical Adaptations to Stress*, Timmermann, B. N., Steelink, C., and Loewus, F. A., Eds., Plenum Press, New York, 1983, 97.
82. **Nobloch, K. H. and Berlin, J.**, Influence of phosphate on the formation of the indole alkaloid and phenolic compounds in cell suspension cultures of *Catharanthus roseus:* I. Comparison of enzyme activities and product accumulation, *Plant Cell Tissue Organ Culture*, 2, 333, 1983.
83. **Vartanian, N.**, Some aspects of structural and functional modifications induced by drought in root systems, *Plant Soil*, 63, 83, 1981.

84. **Kolattukudy, P. E. and Soliday, C. L.,** Effects of stress on the defensive barriers of plants, in *Cellular and Molecular Biology of Plant Stress,* Key, J. L. and Kosuge, T., Eds., Alan R. Liss, New York, 1985, 381.
85. **Amorim, H. V., Dougall, D. K., and Sharp, W. R.,** The effect of carbohydrate and nitrogen concentration on phenol synthesis in Paul's scarlet rose cells grown in tissue culture, *Physiol Plant.,* 39, 91, 1977.
86. **Phillips, R.,** The regulation of synthesis of phenolics in stationary phase cell cultures of *Acer pseudoplatanus* L., *J. Exp. Bot.,* 28, 785, 1977.
87. **Führer, J.,** Early effects of excess cadmium uptake in *Phaseolus vulgaris, Plant Cell Environ.,* 5, 263, 1982.
88. **Stermer, B. A. and Hammerschmidt, R.,** Heat shock induces resistance to *Cladosporium cucumerinum* and enhances peroxidase activity in cucumbers, *Physiol. Plant Pathol.,* 25, 239, 1984.
89. **Walker-Simmons, M., Holländer-Czytko, H., Andersen, J. K., and Ryan, C. A.,** Wound signals in plants: a systemic plant wound signal alters plasma membrane integrity, *Proc. Natl. Acad. Sci. U.S.A.,* 81, 3737, 1984.
90. **Esquerré-Tugayé, M. T., Mazau, D., Pelissier, B., Roby, D., Rumeau, D., and Toppan, A.,** Induction by elicitors and ethylene of proteins associated to the defense of plants, in *Cellular and Molecular Biology of Plant Stress,* Key, J. L. and Kosuge, T., Eds., Alan R. Liss, New York, 1985, 459.
91. **Glass, A. D. M. and Dunlop, J.,** Influence of phenolic acids on ion uptake. IV. Depolarization of membrane potentials, *Plant Physiol.,* 54, 855, 1974.
92. **Mapson, L. W.,** Biosynthesis of ethylene and the ripening of fruit, *Endeavor,* 29, 29, 1970.
93. **Wardale, D. A.,** Effect of phenolic compounds in *Lycopersicon esculeutum* on the synthesis of ethylene, *Phytochemistry,* 12, 1523, 1973.
94. **Yang, S. F.,** Biosynthesis of ethylene, *Arch. Biochem. Biophys.,* 122, 481, 1967.
95. **Itzhaki, H., Borochov, A., and Mayak, S.,** Characterization of an endogenous inhibitor of ethylene biosynthesis in carnation petals, in *Ethylene: Biochemical, Physiological and Applied Aspects,* Fuchs, Y. and Chalutz, E., Eds., M. Nijhoff/W. Junk, The Hague, 1984, 169.
96. **Billot, J.,** Evolution des composés phénoliques au cours de la maturation de la Poire Passe-crassane, *Physiol. Veg.,* 21, 527, 1983.
97. **Vioque, A., Albi, M. A., and Vioque, B.,** Role of IAA-oxidase in the formation of ethylene from 1-aminocyclopropane-1-carboxylic acid, *Phytochemistry,* 20, 1473, 1981.
98. **Hyodo, H., Kuroda, H., and Yang, S. F.,** Induction of phenylalanine ammonia-lyase and increase in phenolics in lettuce leaves in relation to the development of russet spotting caused by ethylene, *Plant Physiol.,* 62, 31, 1978.
99. **Chalker-Scott, L., Fuchigami, L. H., and Harber, R. M.,** Spectrophotometric measurement of leached phenolic compounds as an indicator of freeze damage, submitted to *J. Am. Soc. Hortic. Sci.,* 1988.
100. **Graham, D. and Patterson, B. D.,** Responses of plants to low, non-freezing temperatures: proteins, metabolism, and acclimation, *Annu. Rev. Plant Physiol.,* 33, 347, 1982.
101. **Streltsina, S. A.,** Aspects of the composition of phenolic compounds in the fruits of different species and varietal groups of apple, *Byull. Vses. Inst. Rastenievod.,* 98, 50, 1980.
102. **Hubackova, M.,** Effect of the lignification of grapevine shoots on the frost resistance of buds in winter, *Sb. UVTI, Zahrad.,* 9, 271, 1982.
103. **Kur'yata, V. G.,** Lignification of plants in the autumn-winter period, *Viniti,* 1428, 1978.
104. **Huang, Y. and Wang, Z.,** Cytological determination of cold resistance in fruit trees *(Malus), Acta Hortic.,* 9, 23, 1982.
105. **Sharma, K. P., Sidhu, B. S., and Kanwar, R. S.,** Studies on frost tolerance in sugarcane. III. Some metabolic changes induced by low temperature in the leaf tissue of the spring-planted crop, *J. Res. India,* 18, 109, 1981.
106. **Selstam, E. and Öquist, G.,** Effects of frost hardening of the composition of galactolipids and phospholipids occurring during isolation of chloroplast thylakoids from needles of Scots pine, *Plant Sci.,* 42, 41, 1985.
107. **Bartolo, M. E. and Wallner, S. J.,** Cold hardiness and cellulase resistance induced by wounding, *Plant Physiol.,* Suppl. 80 (Abstr.), 122, 1986.
108. **Fry, S. C.,** Incorporation of [^{14}C]-cinnamate into hydrolase-resistant components of the primary cell wall of spinach, *Phytochemistry,* 23, 59, 1984.
109. **Taiz, L.,** Plant cell expansion: regulation of cell wall mechanical properties, *Annu. Rev. Plant Physiol.,* 35, 585, 1984.

Chapter 6

ETHYLENE-INDUCED STRESS RESISTANCE

R. M. Harber and L. H. Fuchigami

TABLE OF CONTENTS

I. Introduction 82

II. Disease Resistance 82

III. Mechanical Stress Resistance 83

IV. Water Stress Resistance 83

V. Freezing Stress Resistance 84
A. Possible Mechanism 85

VI. Conclusion 86

References 87

I. INTRODUCTION

Ethylene production is a common response to most stresses, including freezing,[1,2] chilling,[3] wounding,[4,5] water,[6,7] salt,[8] and mineral stress,[9,10] as well as disease and insect attack,[11] and exposure to air polutants,[12,13] toxic chemicals,[14] or herbicides.[15] The amount of stress-induced ethylene production is used to quantify the degree of stress a plant has undergone.[16,17] Typically, ethylene production increases with the severity of the stress imposed, up to the point of 50% tissue damage.[1] As the percentage of tissue damage increases over 50%, ethylene production decreases. This is presumably because ethylene formation occurs in living cells, with intact membranes, adjacent to those which are damaged. Ethylene production may also be used as a selection criteria in the breeding of stress resistant varieties. The more tolerant varieties generally produce less ethylene than the sensitive varieties, at a given level of stress.[18,19] Furthermore, it is known that pre-exposure to one stress will often confer resistance to another.[20-23] This knowledge, along with the nearly universal nature of stress ethylene production, suggests ethylene as having the role of a universal, stress-induced defense hormone.

II. DISEASE RESISTANCE

Some of the earliest data suggesting ethylene as being involved in stress resistance comes from pathological studies. Its role in disease resistance has been studied for several years, and has been reviewed by Boller.[24] Thus, it will be only briefly discussed here. Back in the mid-1960s, Clare et al.[25] observed that volatile materials from diseased roots caused an increase in disease resistance of proximal, noninfected roots. Stahmann et al.[26] subsequently showed that the material responsible for this transferal of resistance was ethylene. It was further demonstrated that an exogenous ethylene gas treatment (8 ppm) for 2 days, resulted in increased resistance of sweet potato to infection by *Ceratocystis*.

Two main hypotheses were proposed for the mechanism of this ethylene response. These included the stimulation of peroxidase activity, and the stimulation of photoalexin production.[27] The former hypothesis is partially supported by the fact that ethylene treatment did result in an increase in peroxidase activity. However, a concomitant increase in disease resistance was lacking.[28] Conversely, phytoalexins are known to build up in diseased plants, and to subsequently afford localized protection to the plant in the specific area of infection.[29] Phytoalexin production is also stimulated by ethylene gas treatments.[30,31] Some key enzymes involved in phytoalexin synthesis, such as phenylalanine ammonia lyase (PAL), have been shown to be stimulated following the application of ethylene.[32] However, ethylene had less of a stimulatory effect than many other compounds tested.[27] Paradies et al.[33] have shown ethylene production to be closely correlated with, but not an inducer of phytoalexin production. As summarized by Yang and Pratt,[11] ethylene in some cases does stimulate wound healing, phytoalexin production, or disease resistance, but this is not a general phenomenon. The close correlation between phytoalexin and ethylene production, however, is intriguing and warrants further consideration.

A third alternative for ethylene's involvement in disease resistance is the induction of cell-wall associated, hydroxyproline-rich glycoprotein synthesis. These glycoproteins were positively correlated with pathogen resistance and increased following ethylene application.[34] Inhibitors of ethylene biosynthesis were also shown to prevent the incorporation of ^{14}C-proline into these cell-wall glycoproteins.[35] Also, exogenous 1-aminocyclopropane-1-carboxylic acid (ACC) was found to trigger the synthesis of hydroxyproline-rich glycoproteins.[36]

III. MECHANICAL STRESS RESISTANCE

Ethylene is also implicated in the acquisition of resistance to mechanical perturbation.[23] A common response of plants to mechanical stress is a decrease in stem height and an increase in stem diameter.[23,37,38] These are also typical plant responses to ethylene gas.[39-41] The relationship between ethylene and the thigmomorphogenetic response is strengthened by the observation that inhibition of ethylene biosynthesis prevents the increase in stem thickness following a mechanical stimulus.[23] Furthermore, there is evidence that thigmomorphogenesis is important for the plant's survival under a subsequent and more severe mechanical stress, such as that encountered in a wind storm. For example, a lodging resistant variety of wheat was found to respond to a mechanical stimulus by decreasing its growth, while a lodging susceptible variety showed no response.[42] It has also been shown that when plants are pretreated with a mechanical stress, they are subsequently more resistant to drought or freezing stress.[23] Inhibition of ethylene biosynthesis prevented this mechanically induced hardening to drought, whereas nonstimulated plants treated with ethephon showed an increase in resistance to drought-induced pithiness in tomato plants,[23] but not in celery.[43] In contrast to these reports, Biddington and Dearman[44] found mechanically stressed lettuce seedlings to be less drought tolerant than the controls.

IV. WATER STRESS RESISTANCE

The possibility of ethylene playing a role in a plant's acclimation to water stress has been alluded to several times in the literature,[11,45,46] yet relatively little research has been conducted which specifically addresses this question. El-Beltagy and Hall[45] hypothesized that water stress-induced ethylene may play an adaptive role in the plant's response to stress by accelerating senescence and abscission of its transpiring leaves, thus reducing water loss. Abeles et al.[47] have found that ethylene accelerates abscission of aged leaves more readily than younger leaves. Under drought conditions, it may be especially advantageous to abscise these older leaves on which the stomata may be insensitive to changes in the plant's water relations.[45]

Exogenous ethylene has also been shown to increase the osmolality of cell sap,[48,49] thus suggesting ethylene's role in osmotic adjustment, which is an important mechanism of drought resistance.[50] In contrast, ethylene may also benefit the waterlogged plant by an entirely different mechanism. Ethylene treatments are known to stimulate both hypertrophied lenticel and adventitious root formation,[51-53] which are adaptations for survival in response to flooding.[54]

Further evidence for ethylene's role in water stress acclimation may be found in the comparison of drought-induced ethylene production from resistant and sensitive varieties. Rose[55] found that a drought resistant variety of wheat produced more drought-induced stress ethylene than did a sensitive variety. Also, mildly stressed plants of both varieties increased in stomatal resistance following ethylene treatments. It is tempting to hypothesize that the increased ethylene production in the resistant variety was responsible for its greater drought resistance. This relationship, however, is contrary to other reports which have shown that the more stress sensitive varieties produced more stress ethylene.[18,19] These opposing results may be due simply to differences in severity of the stress imposed. For example, a minimal stress, which is sufficient to induce increased ethylene production in a sensitive variety, may not even be perceived as a stress by the more resistant variety, hence resulting in a smaller increase in ethylene evolution. However, it must also be considered that the absolute amount of ethylene production may not be the most important factor in determining a plant's response to stress. Kapuya and Hall[56] have shown that relative ethylene concentrations of droughted vs. waterlogged plants are not necessarily indicative of resistance in either the

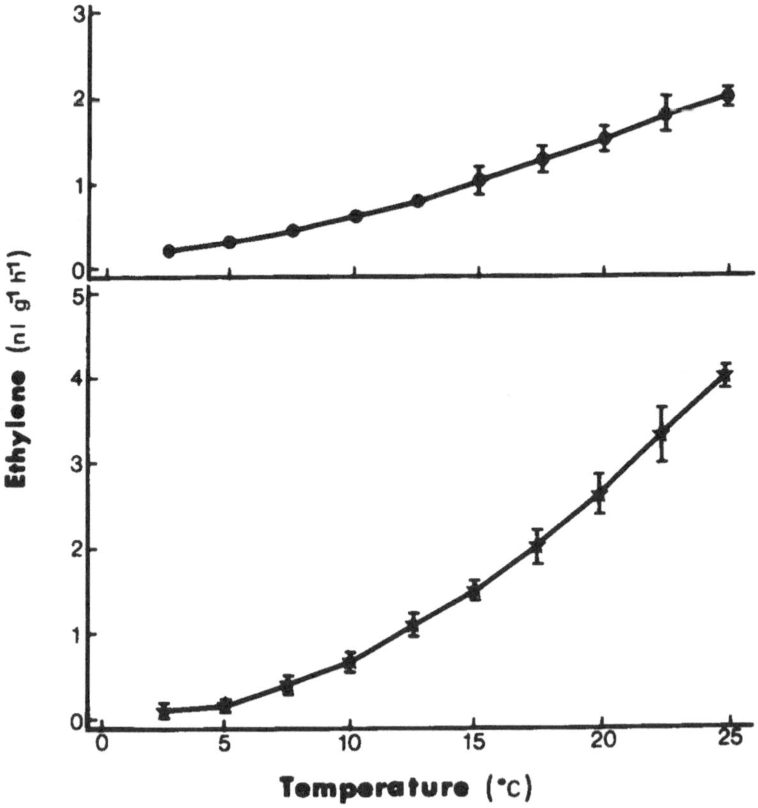

FIGURE 1. Effect of temperature range(2.5 - 25°C) on basal (●) and wound (★) ethylene production. Vertical bars indicate standard error of the mean. (From Field, R. J., *Ann. Bot.*, 47, 215, 1981. With permission.)

xerophyte or the hydrophyte tested. Therefore, a plant's sensitivity to a given level of ethylene must also be taken into account.

V. FREEZING STRESS RESISTANCE

Low temperatures are required for the first stage of cold acclimation in winter cereals,[57] and the second stage in temperate woody plants.[58] The rate of acclimation increases as the temperature decreases.[59,60] A typical acclimating regime for wheat seedlings is 5°C during the day and 2°C at night.[61] Under these conditions, ethylene production in wheat is greatly reduced in comparison to its production at 23°C.[62] This is in agreement with other reports (see review),[63] which have shown ethylene production to decrease in response to lowered temperatures (Figure 1). The rate-limiting step in ethylene production at chilling temperatures is the production of ACC from S-adenosyl methionine.[63] Wang and Adams[3] determined that both ACC and the enzyme involved in its synthesis, ACC synthase, were at reduced levels in tissue held at 2.5°C. Also, the addition of ACC to tissue held at 5°C resulted in increased ethylene production.[64]

Our studies indicate that although ethylene production is low at acclimating temperatures (5°/2°C), the internal level of ethylene can be significantly reduced with the ethylene bio-

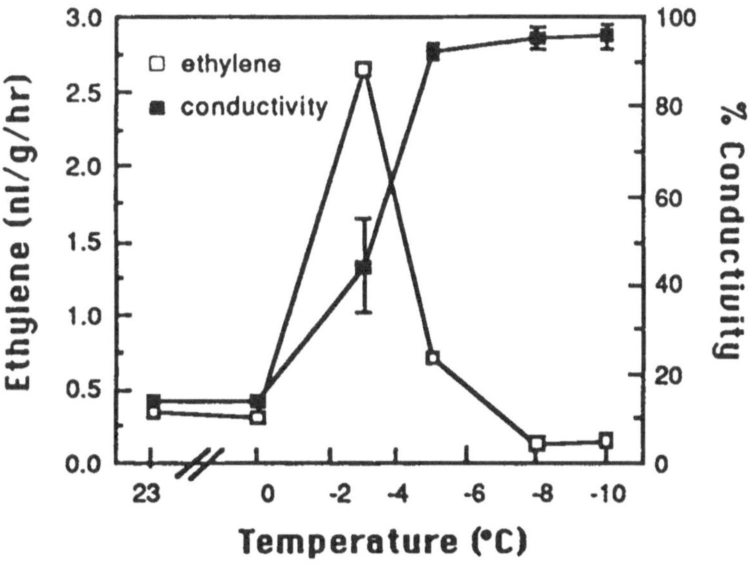

FIGURE 2. Freeze stress-induced ethylene production and percentage of electrolyte leakage from rhododendron "Sappho" leaf disks. (Modified from Harber, R. M. and Fuchigami, L. H., *J. Am. Soc. Hortic. Sci.*, 111, 434, 1986.)

synthesis inhibitor, AVG. There is evidence that the inhibition of ethylene production or its activity during the first week of acclimation, will delay the increase in hardiness of wheat seedlings.[62]

There is also evidence that exogenous ethylene applications will increase the cold hardiness of some plants. Proebsting and Mills[65] found that spring and summer applications of ethephon increased cherry and prune flower bud hardiness to an early winter freeze. Also, mid-October applications of ethephon increased apple shoot cold hardiness during November.[66] Yet, it had no effect on the cold hardiness of citrus seedlings.[67] In our studies, exposing alfalfa suspension cultures to ethylene treatments for 2 days at 23°C, resulted in a 3 to 4°C increase in hardiness. Similarly, wheat seedlings, treated with ethylene during the first week of acclimation at 4°C were hardier than the air-treated controls. If the seedlings were acclimated prior to the ethylene treatment, no effect from the ethylene was observed.[62] Thus, it appears that ethylene is involved at the beginning of the acclimation process.

In woody plants, the rate of acclimation, during the second and third stage of hardiness, is much greater when exposed to subfreezing temperatures as opposed to low, nonfreezing temperatures.[58,68] Coincidentally, sublethal freezing stress results in a dramatic increase in ethylene production upon warming to 23°C (Figure 2). An analogous situation might occur in nature when a sublethal early morning frost is followed by a relatively warm day. Because of this relationship, it was hypothesized that ethylene is the responsible factor for the increased acclimation rate observed in frosted plants. In studies designed to test this theory, we have found that a mild freeze does increase the rate of acclimation in dogwood stem sections, but not in rhododendron leaf discs. Various levels of frost-induced ethylene production were attained by chemically inhibiting its production and by stressing the tissue to a range of sublethal temperatures, with the greater stress yielding the higher ethylene production. There was not, however, a correlation between the amount of stress-induced ethylene production and the rate of acclimation.[62] At this time, the role of ethylene in frost-induced acclimation remains inconclusive.

A. Possible Mechanism

Yoshida[70] has shown the fluidity of the plasma membrane to increase in conjunction with the development of cold hardiness. Increasing the membrane permeability may allow water to pass more freely into the extracellular spaces during freezing, thereby decreasing the likelihood of an intracellular freeze. Alterations of the lipid portion of a cell's membranes is a common occurence during acclimation, and is believed to be necessary for the plant's ultimate resistance to low temperatures.[69,70] Ethylene is known to increase membrane permeability,[71] and it may be doing this by decreasing the phospholipid component.[72,73] Irvine and Osborne[74] have shown that an ethylene treatment of 10 ppm resulted in a 50% decrease in the incorporation of 1-^{14}C glycerol into phospholipids. Also, in both mechanically perturbed and Ethrel-treated plants, the resulting decrease in cell elongation and increase in lateral divisions were correlated with a decrease in free fatty acids and microsomal phospholipids.[75] However, phospholipids are generally found to increase in a plant's membranes, specifically the plasma membrane, during cold acclimation,[69,70,76,77] thus would seem to dispute the contention that ethylene is involved in this manner. Yet, the relation between phospholipids and both ethylene and cold hardiness, although seemingly opposite, suggests that a closer look be taken at their interrelations, especially during the first few days of acclimation.

Ethylene, as previously mentioned, is known to stimulate some key enzymes in photoalexin synthesis, such as PAL and cinnamate 4-hydroxylase.[78] Thus, these enzymes may be responsible for increased production of phenolics other than those commonly referred to as phytoalexins. These phenolics may then be involved in cell wall build-up by forming lignin or suberin. This possibility was discussed in greater depth in the previous chapter.

VI. CONCLUSION

Ethylene has been shown to stimulate many different types of responses in plants, ranging from growth promotion to senescence. Of these responses, several have been implicated in a plant's acclimation to stress. In review, these include (1) retarding stem elongation and increasing diameter, (2) accelerating senescence and abscission, (3) inducing the formation hypertrophied lenticels and adventitious roots, (4) increasing stomatal resistance, (5) increasing osmolality of cell sap, (6) increasing hydroxyproline rich glycoprotein synthesis, and (7) increasing phytoalexin production. The result of ethylene under a given stress is dependent upon the environmental cues, the genetic potential, and the physiological status of the plant. A lowered water potential, decreased oxygen status or reduction in temperature may each allow for slight variations in protein synthesis and enzyme activities, in response to ethylene.

There is a moderate amount of evidence which links ethylene to increased stress resistance in plants, yet its role remains unclear. It might be tempting to conclude that stress-induced ethylene is simply a by-product of the plant's stress response. However, in a evolutionary sense, it would be impractical for a plant to actively synthesize a compound for no reason. This would be especially true under stressful conditions, when the energy supply may be quickly depleted. Of course, ethylene may benefit the stressed plant in ways other than by increasing its resistance to the present or subsequent stress. It may for example, aid in the plant's recovery from stress. There is much research which remains to be done before the role of ethylene in stress resistance can be fully understood.

REFERENCES

1. **Elstner, E. F. and Konze, J. R.,** Effect of point freezing on ethylene and ethane production by sugar beet leaf disks, *Nature (London)*, 263, 351, 1976.
2. **Nissila, P. C.,** Hardiness-maturity relations in red-osier dogwood (*Cornus stolonifera* Michx.), MS thesis, Oregon State University, Corvallis, 1977.
3. **Wang, C. Y. and Adams, D. O.,** Chilling-induced ethylene production in cucumbers (*Cucumis sativus* L.), *Plant Physiol.*, 69, 424, 1982.
4. **Meigh, D. F., Norris, K. H., Craft, C. C., and Lieberman, M.,** Ethylene production by tomato and apple fruits, *Nature (London)*, 186, 902, 1960.
5. **Yu, Y.B. and Yang, S. F.,** Biosynthesis of wound ethylene, *Plant Physiol.*, 66, 281, 1980.
6. **Kobayashi, K., Fuchigami, L. H., and Brainerd, K. E.,** Ethylene and ethane production and electrolyte leakage of water-stressed 'Pixy' plum leaves, *HortScience*, 16, 57, 1981.
7. **Hoffman, N. E., Liu, Y., and Yang, S. F.,** Changes in 1-(malonylamino)cyclopropane-1-carboxylic acid content in wilted wheat leaves in relation to their ethylene production rates and 1-aminocyclopropane-1-carboxylic acid content, *Planta*, 157, 518, 1983.
8. **Garcia, F. G. and Einset, J. W.,** Ethylene and ethane production in 2,4-D treated and salt treated tobacco tissue cultures, *Ann. Bot. (London)*, 51, 287, 1983.
9. **Fuhrer, J.,** Ethylene biosynthesis and cadmium toxicity in leaf tissue of beans (*Phaseolus vulgaris* L), *Plant Physiol.*, 70, 162, 1982.
10. **Matto, A. K., Baker, J. E., and Lieberman, M.,** Copper induced ethylene production by *Spirodela oligorrhiza* is a consequence of singlet oxygen mediated membrane damage, *Plant Physiol.*, 72, 39, 1983.
11. **Yang, S. F. and Pratt, H. K.,** The physiology of ethylene in wounded plant tissue, in *Biochemistry of Wounded Plant Tissues*, Kahl, G., Ed., deGruyter, Berlin, 1978, 596.
12. **Bressan, R. A., LeCureux, L., Wilson, L. G., and Filner, P.,** Emission of ethylene and ethane by leaf tissue exposed to injurious concentrations of sulfur dioxide or bisulfite ion, *Plant Physiol.*, 63, 924, 1979.
13. **Craker, L. E.,** Development of a test tube stress-ethylene bioassay for detecting phytotoxic gases, *Environ. Pollut.*, 28, 265, 1982.
14. **Nee, C. C.,** Overcoming Bud Dormancy with Hydrogen Cyanamide: Timing and Mechanism, Ph.D. thesis, Oregon State University, Corvallis, 1986.
15. **Thompson, C. R., Kats, G., Dawson, P., and Doyle, D.,** Development of protocol for testing effects of toxic substances on plants, EPA Project Sum., 600/S3-81-006, 1981.
16. **Tingey, D. T., Standley, C., and Field, R. W.,** Stress ethylene evolution: a measure of ozone effects on plants, *Atmos. Environ.*, 10, 969, 1976.
17. **Harber, R. M. and Fuchigami, L. H.,** The relationship of ethylene and ethane production to tissue damage in frozen rhododendron leaf disks, *J. Am. Soc. Hortic. Sci.*, 111, 434, 1986.
18. **Chen, Y. Z. and Patterson, B. D.,** Ethylene and 1-aminocyclopropane-1-carboxylic acid as indicators of chilling sensitivity in various plant species, *Aust. J. Plant Physiol.*, 12, 377, 1985.
19. **Corey, K. A., Barker, A. V., and Craker, L. E.,** Ethylene evolution by tomato plants under stress of ammonium toxicity, *Am. Soc. Agron. Abstr.*, 196, 1986.
20. **Rikin, A. and Richmond, A. E.,** Abscisic acid as a common factor in hormonal regulation of resistance to cold and drought in plants, *Isr. J. Bot.*, 24, 54, 1975.
21. **Stout, D. G.,** Alfalfa water status and cold hardiness as influenced by cold acclimation and water stress, *Plant Cell Environ.*, 3, 237, 1980.
22. **Cloutier, Y. and Andrews, C. J.,** Efficiency of cold hardiness induction by dessication stress in four winter cereals, *Plant Physiol.*, 76, 595, 1984.
23. **Jaffe, M. J.,** The involvement of callose and elicitors in ethylene production caused by mechanical perturbation, in *Ethylene: Biochemical, Physiological and Applied Aspects*, Fuchs, Y. and Chalutz, E., Eds., Marinus Nijhoff/Dr. W. Jund Pub., The Netherlands, 1984, 199.
24. **Boller, T.,** Ethylene-induced biochemical defenses against pathogens, in *Plant Growth Substances*, Wareing, P. F., Ed., Academic Press, New York, 1982, 303.
25. **Clare, B., Weber, D. J., and Stahmann, M. A.,** Peroxidase and resistance to *Ceratocystis* in sweet potato increased by volatile materials, *Science*, 153, 62, 1966.
26. **Stahmann, M. A., Clare, B. G., and Woodbury, W.,** Increased disease resistance and enzyme activity induced by ethylene and ethylene production by black rot infected sweet potato tissue, *Plant Physiol.*, 41, 1505, 1966.
27. **Sequeira, L.,** Hormone metabolism in diseased plants, *Annu. Rev., Plant Physiol.*, 24, 353, 1973.
28. **Daly, J. M., Seevers, P. M., and Ludden, P.,** Studies on wheat stem rust resistance controlled at the Sr6 locus. III. Ethylene and disease reaction, *Phytopathology*, 60, 1648, 1970.
29. **West, C. A.,** Fungal elicitors of the phytoalexin response in higher plants, *Naturwissenschaften*, 68, 447, 1981.

30. **Chalutz, E., DeVay, J. E., and Maxie, E. C.,** Ethylene-induced isocoumarin formation in carrot root tissue, *Plant Physiol.*, 44, 235, 1969.
31. **Chalutz, E. and Stahmann, M. A.,** Induction of pisatin by ethylene, *Phytopathology*, 59, 1972, 1969.
32. **Riov, J., Monselise, S. P., and Kahan, R. S.,** Ethylene-controlled induction of phenylalanine ammonia-lyase in citrus fruit peel, *Plant Physiol.*, 44, 631, 1969.
33. **Paradies, I., Konze, J. R., Elstner, E. F., and Paxton, J.,** Ethylene: indicator but not inducer of phytoalexin synthesis in soybean, *Plant Physiol.*, 66, 1106, 1980.
34. **Esquerre-Tugaye, M. T., Lafitte, C., Mazau, D., Toppan, A., and Touze, A.,** Cell surfaces in plant-microorganism interactions. II. Evidence for the accumulation of hydroxyproline-rich glycoproteins in the cell wall of diseased plants as a defense mechanism, *Plant Physiol.*, 64, 320, 1979.
35. **Toppan, A., Roby, D., and Esquerre-Tugaye, M. T.,** Cell surfaces in plant-microorganism interactions. III. *In vivo* effect of ethylene on hydroxyproline-rich glycoprotein accumulation in the cell wall of diseased plants, *Plant Physiol.*, 70, 82, 1982.
36. **Roby, D., Toppan, A., and Esquerre-Tugaye, M. T.,** Cell surfaces in plant-microorganism interactions. V. Elicitors of fungal and of plant origin trigger the synthesis of ethylene and of cell wall hydroxyproline-rich glycoproteins in plants, *Plant Physiol.*, 77, 700, 1985.
37. **Hiraki, Y. and Ota, Y.,** The relationship between growth inhibition and ethylene production by mechanical stimulation in *Lilium longiflorum, Plant Cell Physiol.*, 16, 185, 1975.
38. **Biddington, N. L.,** The effects of mechanically-induced stress in plants — a review, *Plant Growth Reg.*, 4, 103, 1986.
39. **Eisinger, W. R. and Burg, S. P.,** Ethylene-induced pea internode swelling. Its relation to ribonucleic acid metabolism, wall protein synthesis, and cell wall structure, *Plant Physiol.*, 50, 510, 1972.
40. **Poovaiah, B. W.,** Promotion of radial growth by 2-chloroethylphosphonic acid in bean, *Bot. Gaz. (Chicago)*, 135, 289, 1974.
41. **Robitaille, H. A. and Leopold, A. C.,** Ethylene and the regulation of apple stem growth under stress, *Physiol. Plant.*, 32, 301, 1974.
42. **Steucek, G. L. and Gordon, L. K.,** Response of wheat *(Triticum aestivum)* seedlings to mechanical stress, *Bot. Gaz. (Chicago)*, 136, 17, 1975.
43. **Pressman, E., Huberman, M., Aloni, B., and Jaffe, M. J.,** Pithiness in plants. I. The effect of mechanical perturbation and the involvement of ethylene in petiole pithiness in celery, *Plant Cell Physiol.*, 25, 891, 1984.
44. **Biddington, N. L. and Dearman, A. S.,** The effects of mechanically-induced stress on water loss and drought resistance in lettuce, cauliflower and celery seedlings, *Ann. Bot. (London)*, 56, 795, 1985.
45. **El-Beltagy, A. S. and Hall, M. S.,** Effect of water stress upon endogenous ethylene levels in *Vicia faba*, *New Phytol.*, 73, 47, 1974.
46. **Leiberman,M.,** Biosynthesis and action of ethylene, *Annu. Rev. Plant Physiol.*, 30, 533, 1979.
47. **Abeles, F. B., Holm, R. E., and Gahagan, H. E.,** Abscission: the role of ageing, *Plant Physiol*, 42, 1351, 1967.
48. **Eisinger, W., Croner, L. J., and Taiz, L.,** Ethylene-induced lateral expansion in etiolated pea stems. Kinetics, cell wall synthesis, and osmotic potential, *Plant Physiol.*, 73, 407, 1983.
49. **Hayashi, T. and Maclachlan, G.,** Pea xyloglucan and cellulose. III. Metabolism during lateral expansion of pea epicotyl cells, *Plant Physiol.*, 76, 739, 1984.
50. **Boyer, J. S.,** Plant productivity and environment, *Science*, 218, 443, 1982.
51. **Abeles, F. B.,** *Ethylene in Plant Biology*, Academic Press, New York, 1973, chap. 6.
52. **Doubt, S. L.,** The response of plants to illuminating gas, *Bot. Gaz. (Chicago)*, 63, 209, 1917.
53. **Woffenden, L. M. and Priestley, J. H.,** The toxic action of coal gas upon plants. II. The effect of coal gas upon cork and lenticel formation, *Ann. Appl. Bot.*, 11, 42, 1924.
54. **Tang, Z. C. and Kozlowski, T. T.,** Physiological, morphological and growth responses of *Platanus occidentalis* seedlings to flooding, *Plant Soil*, 66, 243, 1982.
55. **Rose, E.,** Water relations of winter wheat, PhD thesis, Kansas State University, Manhattan, 1985.
56. **Kapuya, J. A. and Hall, M. A.,** Plant sensitivity to endogenous ethylene in relation to species characteristics, *Z. Pflanzenphysiol.*, 113, 461, 1984.
57. **Trunova, T. I.,** Mechanism of winter wheat hardening at low temperature, in *Plant Cold Hardiness and Freezing Stress*, Li, P. H. and Sakai, A., Eds., Academic Press, New York, 1982, 41.
58. **Weiser, C. J.,** Cold resistance and injury in woody plants, *Science*, 169, 1269, 1970.
59. **Gusta, L. V., Fowler, D. B., and Tyler, N. J.,** Factors influencing hardening and survival in winter wheat, in *Plant Cold Hardiness and Freezing Stress*, Li, P. H. and Sakai, A., Eds., Academic Press, New York, 1982, 23.
60. **Fuchigami, L. H., Weiser, C. J., Kobayashi, K., Timmis, R., and Gusta, L. V.,** A degree growth stage (°GS) model and cold acclimation in temperature woody plants, in *Plant Cold Hardines and Freezing Stress*, Li, P. H. and Sakai, A., Eds., Academic Press, New York, 1982, 93.

61. **Tanino, K. K. and McKersie, B. D.,** Injury within the crown of winter wheat seedlings after freezing and icing stress, *Can. J. Bot.*, 63, 432, 1985.
62. **Harber, R. M.,** unpublished data, 1986.
63. **Field, R. J.,** The effect of temperature on ethylene production by plant tissues, in *Ethylene and Plant Development*, Roberts, J. A. and Tucker, G. A., Eds., Butterworths, London, 1985, 47.
64. **Field, R. J.,** The effect of low temperature on ethylene production by leaf tissue of *Phaseolus vulgaris* L., *Ann. Bot. (London)*, 47, 215, 1981.
65. **Proebsting, E. L., Jr., and Mills, H. H.,** Effects of growth regulators on fruit bud hardiness in *Prunus*, *HortScience*, 4, 254, 1969.
66. **Raese, J. T.,** Induction of cold hardiness in apple tree shoots with ethephon, NAA, and growth retardants, *J. Am. Soc. Hortic. Sci.*, 102, 789, 1977.
67. **Young, R.,** Effect of growth regulators on citrus seedling cold hardiness, *J. Am. Soc. Hortic. Sci.*, 96, 708, 1971.
68. **Sakai, A.,** Studies of frost hardiness in woody plants, II. Effect of temperature on hardening, *Plant Physiol.*, 41, 353, 1966.
69. **Yoshida, S.,** Chemical and biophysical changes in the plasma membrane during cold acclimation of mulberry bark cells *(Morus bombycis* Koidz. cv. Goroji), *Plant Physiol.*, 76, 257, 1984.
70. **Sikorska, E. and Kacperska-Palacz, A.,** Phospholipid involvement in frost tolerance, *Physiol. Plant.*, 47, 144, 1979.
71. **Lyons, J. M. and Pratt, H. K.,** An effect of ethylene on swelling of isolated mitochrondria, *Arch. Biochem. Biophys.*, 104, 318, 1964.
72. **Simon, E. W.,** Phospholipids and plant membrane permeability, *New Phytol.*, 73, 377, 1974.
73. **Koehler, D. E. and Lewis, L. N.,** Effect of ethylene on plasma membrane density in kidney bean abscission zones, *Plant Physiol.*, 63, 677, 1979.
74. **Irvine, R. F. and Osborne, D. J.,** The effect of ethylene on glycerol incorporation into phospholipids of etiolated pea stems, *Biochem. J.*, 136, 1133, 1973.
75. **Erner, Y. and Jaffe, M. J.,** Thigmomorphogenesis: membrane lipid and protein changes in bean plants as affected by mechanical perturbation and Ethrel, *Physiol. Plant.*, 58, 197, 1983.
76. **de la Roche, I. A., Andrews, C. J., Pomeroy, M. K., Weinberger, P., and Kates, M.,** Lipid changes in winter wheat seedlings *(Triticum aestivum)* at temperatures inducing cold hardiness, *Can. J. Bot.*, 50, 2401, 1972.
77. **Horvath, I., Vigh, L., Belea, A., and Farkas, T.,** Hardiness dependent accumulation of phospholipids in leaves of wheat cultivars, *Physiol. Plant.*, 49, 117, 1980.
78. **Rhodes, M. J. C. and Woolorton, L. S. C.,** Phenolic metabolism in fruit and vegetable tissue under stress, *ARC Food Reseasrch Institute, Biennial Report, 1977-1978,* Norwich, UK.

Chapter 7

ENVIRONMENTAL CONTROL OF GROWTH BEHAVIOR AND COLD HARDINESS IN ARCTIC AND SUBARCTIC PLANTS

Olavi Junttila and Åse Kaurin

TABLE OF CONTENTS

I. Introduction 92

II. Perennial Grass Species 92
A. Growth Behavior 92
B. Cold Hardiness 94

III. Cloudberry (*Rubus chamaemorus* L.) 95
A. Growth Behavior 95
B. Cold Hardiness 96

IV. Woody Plants (*Salix*) 98
A. Growth Behavior 98
B. Cold Hardiness 100

V. Concluding Remarks 104

References 104

I. INTRODUCTION

There are no exact and generally accepted definitions for the northern areas, but they will include the Arctic, the Subarctic, and (at least parts of) the boreal zone.[1] The Arctic can be considered to cover areas north of the tree line, and the Subarctic to consist of areas between the tree species line and the economic tree line, but definitions of these concepts are greatly discussed.[2] Climatically the northern areas are characterized by large annual and relatively small daily fluctuations. The growing season is short and cold. In the Arctic the accumulated heat sum (degree days above 0°C) is less than 600. Except for the coastal areas, especially those in the Fennoscandia, winter temperatures are extremely low. A typical feature for these areas is the drastic annual rhythm in the light conditions. In the Arctic there is a 24-hr photoperiod during the whole growing season. Due to the low sun elevation, plants growing in these areas are often exposed for longer periods to special light qualities with increased relative proportions of irradiation in the blue and the far-red regions.[3] At 70° N latitude the period with midnight sun is about 70 days, and even at the end of the growing season, plants are growing under long day conditions. Due to these extreme growth conditions plants in the northern areas are interesting objects for studies on adaptation.

Larcher and Bauer have given a comprehensive discussion of ecological aspects of resistance to low temperature and they also referred to several studies made on arctic and boreal plant species.[4] Physiological and ecological aspects of cold hardiness in boreal conifers have recently been discussed by Sakai and Eiga.[5] In this chapter we will present mainly results from studies conducted at the University of Tromsø. In these studies, both the climatic regulation of growth and the development of cold hardiness in herbaceous (grass species, cloudberry) and woody (*Salix*) plants have been investigated.

II. PERENNIAL GRASS SPECIES

A. Growth Behavior

Environmental control of growth and development in northern grass species has recently been reviewed by Heide.[6] Both the growth habit and the total dry matter production of northern grass species are affected by temperature and photoperiod. Leaf elongation, as well as stem elongation, is generally promoted by a long photoperiod, and the critical photoperiod for leaf elongation is longer in northern than in southern ecotypes.[7] Endal has demonstrated that in arctic ecotypes (from 78° N latitude) of *Deschampsia alpina* and *Festuca vivipara* the maximum leaf elongation is obtained under 24-hr photoperiods (Figure 1).[8] In contrast to what is the case with shoot elongation in woody plants, the critical photoperiod for leaf elongation in grass is greatly influenced by temperature. For example, in northern ecotypes of *Poa pratensis* (from 71 to 69° N latitude) the critical photoperiod for leaf growth was about 18, 16, and 14 hr at 8°, 13°, 8°, and 13°C, respectively.[7] In a southern ecotype (from 58° N latitude) the effect of daylength on leaf elongation was significant at the lowest temperature (8°C) only.[7] Also, the results presented in Figure 1 clearly show an interaction between photoperiod and temperature in regulation of leaf growth in northern grass species. Differences in critical photoperiods between various latitudinal ecotypes are smaller in grass species than in woody plants. In addition to greatly reduced leaf elongation, short photoperiods, especially when combined with low temperature, also induces a prostrate and rosetted growth habit in northern grass species.

Some recent studies by Heide and his co-workers have clearly established that the dry matter production in northern grass species is significantly enhanced by long photoperiods.[9-12] This is a real photomorphogenetic effect and can be obtained by using low irradiance light during the daylength extension. According to these studies, continuous light can compensate for large differences in temperature (up to 6°C). Similar responses have

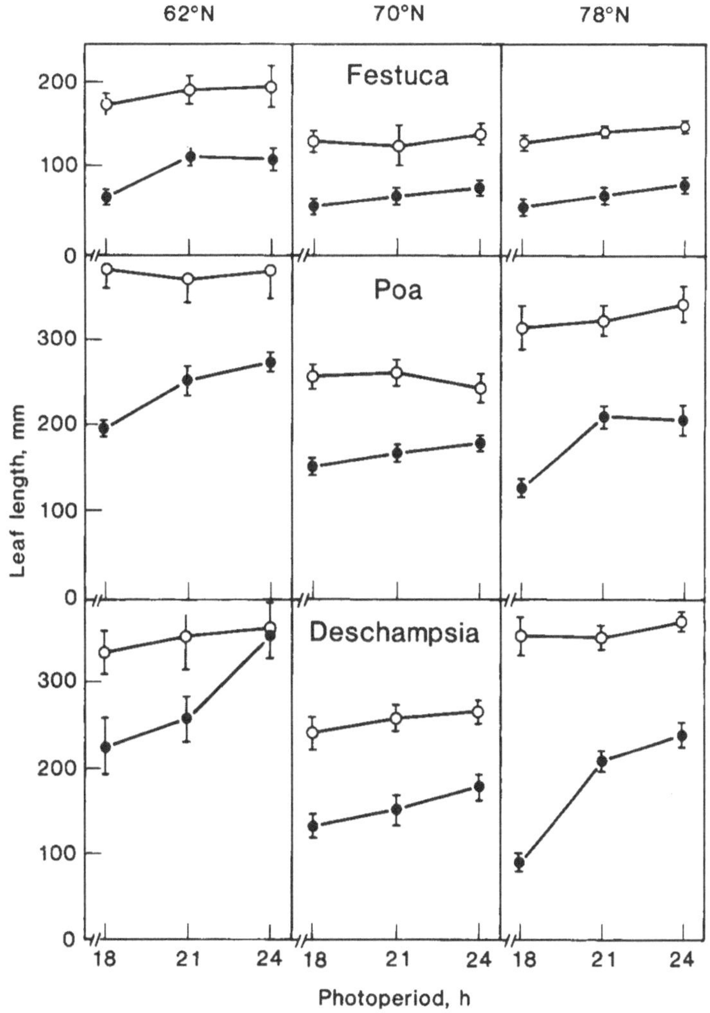

FIGURE 1. Interaction of temperature and photoperiod on leaf elongation in three ecotypes of *Festuca vivipara*, *Poa alpina var. alpina*, and *Deschampsia alpina*. Symbols: ● 12/7°C, ○ 15/12°C (12/12 hr).[8]

earlier been reported by Eagles and Østgård,[13] and according to Larsen and Kaurin the dry matter production in *Dactylis glomerata* was more than doubled when the photoperiod was extended from 8 to 24 hr.[14] In the same experiment they found that the relative content of total carbohydrates in *D. glomerata* increased with decreasing temperature and with increasing photoperiods.

Also, field experiments conducted at 70° N have shown positive effects of long photoperiods for dry matter production.[15] A great yield reduction was observed in three grasses by shortening the natural 24 hr photoperiod. This reduction was most pronounced in the northern varieties. Contrary to controlled environment experiments, an increased leaf length and leaf width at shortened photoperiods was observed. Tiller number was reduced, and the growth was more erect with more green at short days.[15]

Many northern grass species have a dual induction requirement for flowering: short photoperiods and/or temperature are needed for the primary induction, and long photoperiod is necessary for culm development and anthesis.[6] In high latitude ecotypes of *P. pratensis* low temperature and short days are interchangeable for primary induction.[9,16] Inflorescence primordia in these ecotypes are developed early in the fall, and flowering occurs usually quite soon after snow melt. The northern ecotypes seem to require longer photoperiods for culm growth and anthesis than the southern ecotypes.[6,16] According to Endal, arctic ecotypes (78°N) of *Deschampsia, Festuca,* and *Poa* have a low optimum temperature for culm development.[8] For example, temperatures above 9°C reduced flowering in *Festuca.*

Vivipary, or proliferation of the spikelets, is a prominent feature of many arctic (and alpine) grass species and may be an adaptive response to the environmental stresses in these adverse conditions.[17] Vivipary is in some cases a genetic trait for the species, but it can also be affected by photoperiod and temperature.[18] Viviparous plantlets are an effective way of vegetative propagation. These plantlets are developed during the summer and some of them will develop roots before the winter. Viviparous plantlets must have a high potential for hardening. These structures are well suited for studies of cold hardiness, and some results from such investigations are reported here.

B. Cold Hardiness

In northern Norway, the grassland occupies about 90% of the cultivated farm land. Winter damage, causing great variations in the yields, can strongly influence farm economy, and the need for suitable varieties is therefore obvious. During a period of 13 years, trials for hardiness determination of different timothy (*Phleum pratense*) strains have been conducted on freeland in Tromsø.[19] These trials have shown that strains from northern Norway were the most hardy ones. Strains from the southern parts of Scandinavia were strongly thinned during the first winter while those from Scotland, Alaska, and Canada were totally eliminated.

Winter damage in perennial grass species can be caused by several reasons, both abiotic (icing, low temperature, etc.) and biotic (pathogen fungi) ones. In *P. pratensis* and *Festuca pratensis* there was no correlation between cold hardiness and ice tolerance while there was a significant positive correlation between cold hardiness and resistance to fungal infection in *P. pratensis* hardened at 1°C.[20,21] Differences between selected cultivars in resistance to fungi may be caused by different potentials for cold hardening under inappropriate hardening conditions.[22]

It has been shown that freezing tolerance of grasses may be improved by selection based on artificial freezing tests.[23-25] In *D. glomerata* selection for high freezing tolerance during three generations clearly increased the tolerance in one cultivar from 55°N as well as in one from 65°N (Figure 2).[25]

Hardening in northern grass species is affected by temperature and photoperiod. Induction of cold hardiness by low temperature is well documented, but the role of photoperiod is still an open question. For induction of cold hardening the temperature must be below a certain threshold.[26,27] Below this threshold (6°C) photoperiod seems to have little effect on hardening in *D. glomerata* from 65°N, while hardening will decrease at prolonged photoperiods (>16 hr) above this threshold temperature, especially in less hardy varieties.[14,28] Lawrence et al. [29] found that for the same daily energy input, photoperiods of 8 hr resulted in higher tolerance than 16 hr in *Lolium perenne*. Larsen showed that plants of *D. glomerata* were less freezing tolerant when grown at long compared to short photoperiods.[28] This was also shown in *P. pratensis*[7] while the opposite was observed in viviparous plantlets of *P. alpina*.[30] In *D. glomerata* from 65°N, shortening the photoperiod from 16 to 10 hr for 4 to 14 days before hardening significantly increased the hardening ability.[28]

Results from several studies indicate correlations between water content and/or accumulation of carbohydrates and cold hardiness in northern grass species. During acclimation in fall there is a decrease in water content and an increase in the content of carbohydrates. This is a reaction to temperature as well as to photoperiod. During hardening at 1°C under 16-hr photoperiod, percent dry matter decreased in plants of *D. glomerata* (grown at low temperature (6°C) and short photoperiod (8 hr) before hardening). Percent dry matter in-

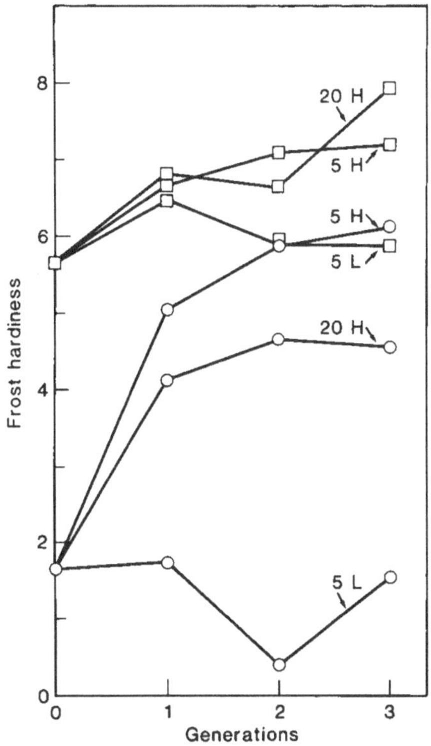

FIGURE 2. Response to selection for freezing tolerance in two cultivars of *Dactylis glomerata*. The tolerance is scored from 0 (dead) to 9 (no damage). (□) "Hattfjelldal", (○) "Unke". 20 and 5 hr: 20 and 5% selection in high direction, 5 L: 5% selection in low direction.[25]

creased most in plants grown at 16-hr photoperiod.[14] The concentration of carbohydrates increased significantly, and the cold hardiness increased with about 70% during the hardening treatment.

P. alpina grown under long day conditions was hardier (LT_{50} = −35°C) and had more carbohydrates than plants grown under short day conditions(LT_{50} = −26°C).[30] When plants were exposed to monospectral blue, red or far red light for 15, 60, or 120 min at the end of the photoperiod, there was no consistent correlation between cold hardiness and carbohydrate concentration.[30]

Reducing the water content in *P. alpina* using polyethylene glycol (PEG)-6000 could, to some extent, eliminate the need for cold hardening treatments. The cold hardiness increased from −8°C (=LT_{50}) to −14°C when the water content was reduced from 60 to 48%. If the temperature during exposure to PEG-6000 was reduced from 18° to 6°C, all plants survived −14°C.[31] This increase in cold hardiness is, however, much less than the level of hardiness (−40°C) obtained using 3 weeks treatment at −3°C.

III. CLOUDBERRY (*Rubus chamaemorus* L.)

A. Growth Behavior

Cloudberry is a circumpolar arctic and subarctic species. It is a common perennial small fruit on oligotrophic mires; in some cases it may even be the dominating species. Berries of this dioecious plant are highly valued in Scandinavia, and cloudberry is of economic importance in the northern areas. It produces underground rhizomes, and during the summer

Table 1
NUMBER OF DAYS FROM TERMINATION OF CHILLING TO BUDBREAK IN 50% OF THE BUDS OF VARIOUS CLOUDBERRY ECOTYPES AS AFFECTED BY TEMPERATURE OF THE PREVIOUS SEASON

	Days to 50% budbreak (year n + 1) origin of the ecotype, °N lat.						
Temperature °C (year n)	**54**	**58**	**63**	**67**	**70**	**78**	**Mean**
9	12	12	8	8	8	6	9
15	16	16	11	10	12	8	12
21	18	18	13	12	15	10	14
Mean	15	15	11	10	12	8	

Note: Temperature year n + 1: 15°C.[32]

vegetative and generative winterbuds are developing on the shoot basis and on the rhizomes. These buds are dormant and thus have a chilling requirement for budbreak. The dormancy is probably not very deep because 25% of the buds were able to break after 30 days between 0 and 3°C.[32] The optimum chilling temperature is close to 0°C.

In the spring, after natural chilling treatment out-of-doors, the budbreak is only slightly affected by growth temperature.[33] However, high temperature during bud development (preceding season) seems to delay the bud break (Table 1). Results presented in Table 1 show further that the northern ecotypes of cloudberry start growing significantly earlier than the southern ones.[32] The same is true with flowering. It should be mentioned that even the northernmost ecotypes of cloudberry are able to develop flowerbuds under both long day and short day conditions, and in this respect no specific adaptation to the northern light conditions has been found.[32]

Elongation growth of cloudberry shoots developing from the winterbud is partly predetermined. The elongation period is just 2 to 3 weeks long and it is only slightly affected by the growth conditions. The rate of shoot growth is increased by long days (24 hr) and high temperature (18°C) compared to short days (12 hr) and low temperature (12°C).[34] Also the number of winterbuds developing on a plant is affected positively by long photoperiod and high temperature conditions.[34]

B. Cold Hardiness

Survival of cloudberry is dependent on a successful overwintering of the winterbuds and the rhizomes. These organs are just at the soil surface or slightly below the surface, and normally they are covered by snow during most of the winter. Like other similar plants, cloudberry has a limited capacity for hardening: approximately −20° to −25°C.[34-37] The annual variation of hardiness is, though, quite clear as shown in Figure 3.[36,37] The results presented in Figure 3 also indicate a positive relationship between frost hardiness and the amount of soluble carbohydrates. Increase of cold hardiness in the fall is, most probably, induced by decreasing temperature. No significant effect of photoperiod on hardening of cloudberry could be detected in an experiment conducted under controlled conditions (Figure 4).[34] However, the effect of temperature was marked, and exposure to freezing temperature (−4°C) was required to develop maximum hardiness. These responses resemble those ob-

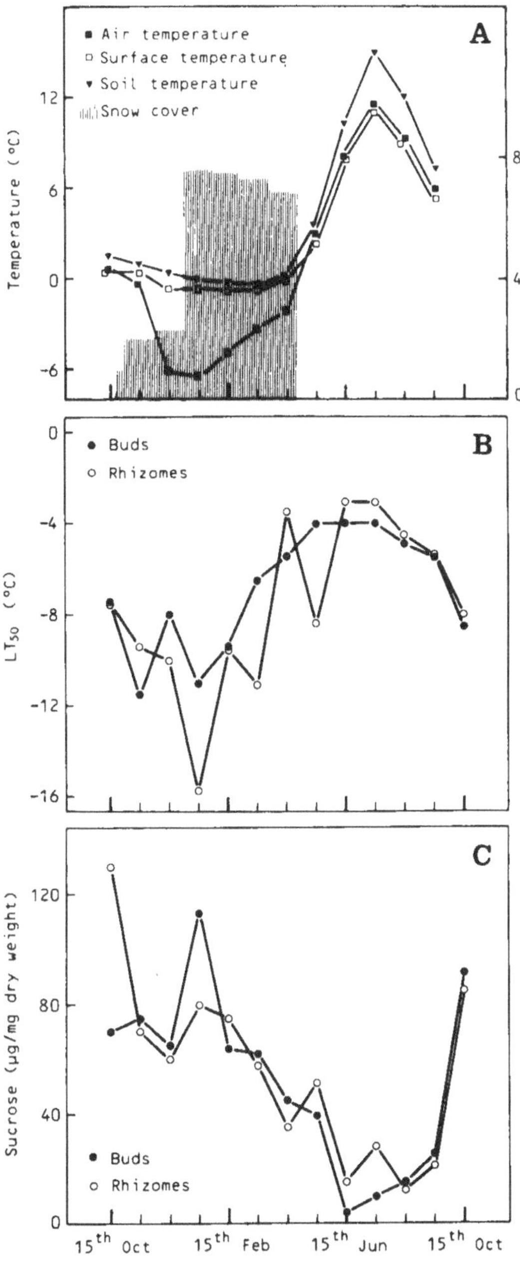

FIGURE 3. Temperatures, LT_{50} and sucrose content from October 1978 to October 1979. (A). Monthly means of snow depth and approximate minimum temperatures for air, soil surface, and soil (10 cm depth). (B). Frost hardiness based on LT_{50} in buds and rhizomes of cloudberry. Each point represents the means of 40 buds or 30 rhizome pieces. (C). Sucrose content in buds and rhizomes of cloudberry.[36]

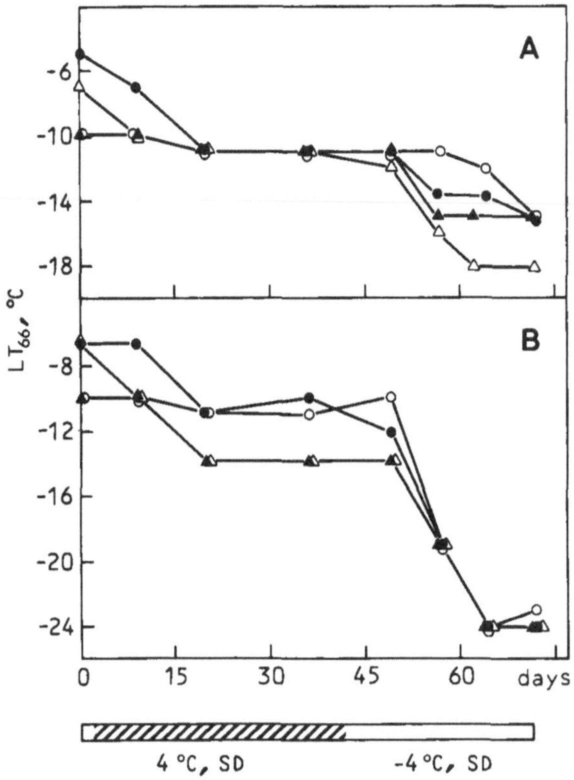

FIGURE 4. Frost hardiness in cloudberry vascular connective tissue (A) and buds (B), 1976. Symbols: △ 18°C-24 hr photoperiod; ▲ 18°C-24-hr photoperiod; ○ 12°C-24-hr photoperiod; ● 12°C-12-hr photoperiod. Vertical bars denote SE (n = 20).[34]

tained with winter cereals.[38-40] No studies have been made to investigate the possible effects of light quality on the develoment of frost hardiness in cloudberry.

Under natural conditions the dehardening in cloudberry was found to start in February (Figure 3). At that time, the plants were still covered with a thick layer of snow and temperature conditions for the plants were quite stable, close to 0°C. These results suggest that the dehardening in cloudberry is at least partly controlled by an inner rhythm.[36]

Development of hardiness in various extreme ecotypes of cloudberry has not been compared in any studies. Growth studies have shown differences between ecotypes with respect to the onset of growth in the spring, and this could indicate that there are differences in the dehardening process. No significant differences in the midwinter hardiness could be detected between northern cloudberry populations from maritime and continental localities.[35] Wintering strategy of cloudberry is obviously based on the protection by stable snow cover, and it has not been necessary for the species to develop extreme hardiness. However, it would be of interest to make a comparative investigation with geographically extreme ecotypes of cloudberry.

IV. WOODY PLANTS (*Salix*)

A. Growth Behavior

Cessation of apical growth is generally a prerequisite for cold hardening in northern species of woody plants. In seedlings of most species this process is primarily regulated by pho-

Table 2
BUD-SET AND COLD HARDINESS IN SEEDLINGS OF *S. PENTANDRA* (ECOTYPE FROM 70°N LATITUDE) AS AFFECTED BY LIGHT QUALITY AND TEMPERATURE

			LT_{50} (°C) for buds after	
Light treatment	**R/FR ratio**	**Bud-set (%)**	**3 weeks at 18°C**	**3 weeks at 18°C + 3 weeks at 3°C**
Darkness (SD)	—	100	< −12	−25
Incandescent (I)	0.8	14	−4	−10
Warm white (ww)	3.1	69	−6	−8
Ww + blue light	3.4	95	−11	−11
Cool white (cw)	3.2	100	−11	−17
Cw + I	1.4	2	> −3	−7
Red	24.7	100	−9	−17
				($LSD_{0.05}$: 2.6)

Note: All plants got photosynthetic light (fluorescent tubes and incandescent lamps, 2200 μW cm^{-2}) for 12 hr and the indicated light qualities (about 200 μW cm^{-2}) were given for additional 12 hr. LT_{50}: lethal temperature for 50% of the buds, R: red, FR: far-red. Light treatments were given also during the low temperature period. ww and cw: warm and cool white fluorescent tubes.

toperiod and it has been shown by several studies that there exists a clinical variation in critical photoperiods for cessation of shoot growth with latitude and altitude.[41-45] The higher the latitude, or the altitude, the longer is the critical photoperiod for apical growth cessation. In birch *(Betula pubescens)* and willow *(Salix pentandra)* ecotypes from 70°N latitude, the critical photoperiod is about 22 hr and this daylength is passed during the first week of August. The critical photoperiod for cessation of elongation growth seems to be quite constant over a range of temperature treatments (12 to 24°C),[43] but also temperature has an influence on growth cessation.[43,45] Low "night" temperature combined with a higher day temperature induces apical growth cessation both in spruce *(Picea abies)* and in willow even under 24-hr photoperiods.[43,45,56] The northern ecotype of *S. pentandra* (70°N) was found to be more sensitive for this low temperature treatment than the southern ecotype (60°N).[45,46] In addition, both the growth rate and the rate of response of the plant to photoperiod is dependent on temperature. Consequently, the regulation of growth cessation can be described as a joint effect of heat sum and photoperiod.[47]

Northern ecotypes of woody plants are less sensitive to light than corresponding southern ecotypes.[46,48,49] For example, daylength extension with incandescent lamps giving 0.4 mW m^{-2} nm^{-1} at 610 nm was enough to give a long day effect in the southern ecotype of *S. pentandra* (60°N), while at least 10 times higher irradiance was needed to give continuous growth in a northern ecotype (70°N).[49] In addition to this difference in light sensitivity, the northern ecotypes also seem to have developed specific reactions to light quality (spectral energy distribution). Results presented in Table 2 clearly show that the northern ecotype of *S. pentandra* requires a high proportion of far-red irradiation during the daylength extension to give a long day effect. The results also indicate that blue light can partly induce apical growth cessation. No such effects could be shown in a southern ecotype of *S. pentandra.*[49] It should also be mentioned that ecotypes of *S. pentandra* from 70°N latitude did not respond to a night break treatment.[50] It is possible that these types of adaptations to light climate are common not only to northern ecotypes but also to several northern species.

The sensitivity to photoperiod in woody plants decreases with the age, and the cessation of apical elongation growth in shoots of adult *S. pentandra* trees occurs more or less

Table 3
TIME FOR APICAL GROWTH CESSATION IN 50% OF THE SEEDLINGS AND WINTER DAMAGE EXPRESSED AS PERCENTAGE OF SHOOT LENGTH (IN PARENTHESIS) IN SEEDLINGS OF *S. PENTANDRA* GROWN OUT-OF-DOORS AT TROMSØ, 70°N LATITUDE

Ecotype, cross	Critical photop. (hr)	Number of days after August 1—(% damage)		
		1980 (1980/81)	1981 (1981/82)	1982 (1982/83)
70°N	22	16 (7.2)	32 (5.9)	16 (0.3)
64°N	20	60(13.7)	46 (6.7)	26 (4.2)
60°N	16	69(34.6)	58(10.5)	49(11.7)
70° × 60°N	19	64(13.4)	48 (8.8)	35 (5.9)

Note: The critical photoperiod has been determined in phytotron experiments.

independently of the photoperiod.[50] Experiments with adult plants of *S. polaris* have shown that shedding of leaves in an ecotype from Spitsbergen (78°N) was not affected by photoperiod when the temperature was 6°C, but at 15°C continuous light delayed the shedding of leaves.[51] In an ecotype from 70°N latitude leaf shedding at 6°C was significantly enchanced by short photoperiod (less than 18 hr) compared to long photoperiod.[51] Under natural conditions in Spitsbergen *S. polaris* starts shedding the leaves in the beginning of August when the daylength still is 24 hr.[52]

When various latitudinal ecotypes of *S. pentandra* are grown out-of-doors at a high latitude, the timing of apical growth cessation and subsequent shedding of leaves are clearly related to the photoperiodic reactions of the ecotype (Table 3). Differences between various years are supposed to be partly due to differences in temperature conditions, partly due to the aging process.

B. Cold Hardiness

Cold hardiness in adult trees of *S. pentandra* increases rapidly during the fall, and in midwinter twigs will survive −196°C.[37] In this study the hardiness cycle during the winter in several species growing both at a maritime and at a continental site were compared, and the results indicated that in six species, *(Betula pubescens, Picea abies, Salix caprea, Sorbus aucuparia, Vaccinium myrtillus, V. vitis-idaea)* the hardening developed faster at the continental than at the coastal location. No such difference could be shown in *S. pentandra*.

The early phases of hardening in *S. pentandra* have been studied further in some recent experiments. Samples were collected (1985 and 1986) from field grown plants of various ecotypes and tested for cold hardiness. Plants were cut down in the spring and this resulted in a vigorous growth and later cessation of apical growth (Figure 5) than in undisturbed plants (see Table 3). However, differences between the studied ecotypes were similar to those found earlier; the northern ecotype (from 70°N latitude) stopped growing 5 weeks earlier than the southern one (from 60°N latitude). At October 4 the southern ecotype was significantly less hardy (LT_{50} for buds = −4.5°C) than the northern ecotype (LT_{50} for buds = −15°C). Both an ecotype from 64°N latitude and the F_1 population of a cross between the southern and the northern ecotype had an intermediate hardiness level. Cold hardiness of the northern ecotype increased much faster than indicated in Figure 5 (the test temperatures used were not low enough to detect the LT_{50} limit), but in the other ecotypes hardening developed at a similar rate. By November 13 all ecotypes had obtained hardiness to survive,

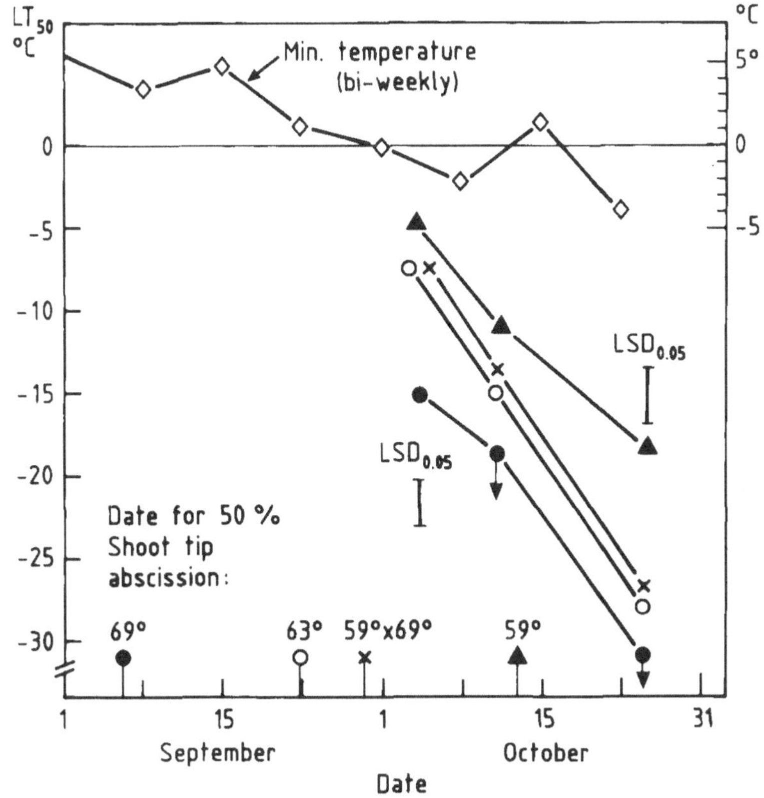

FIGURE 5. Dates for 50% shoot tip abscission and development of cold hardiness of the buds in various ecotypes of *Salix pentandra* grown out-of-doors at 70°N latitude. $LSD_{0.05}$ on October 4 is calculated for all ecotypes. The ecotype from 70°N is not included in calculation of $LSD_{0.05}$ on October 27.

partly or completely, an exposure to liquid nitrogen (after freezing down to −45°C). All the buds of the northern ecotype had a 100% survival. The percentage of survival was 30% for the topmost (0 to 10 cm) and 69% for the more basal (10 to 30 cm) buds of the southern ecotype.

This marked difference between the northern ecotype and the other ecotypes was clearly confirmed in 1986 when the studies were repeated. For example, October 17 the ecotype from 70°N latitude showed no damage when frozen down to −50°C, while the other ecotypes had the following LT_{50} values: 60°N: −21°C; 64°N: −25°C; cross 60° × 70°N: −24°C. Also these results were obtained with samples from plants which were cut down in the spring. Cessation of apical growth in these plants occurred one to two weeks later than in undisturbed plants. For example, undisturbed and cut-down plants of the ecotype from 60°N latitude reached 50% growth cessation September 15 and September 27, respectively. Consequently, freezing down to −20°C of samples collected October 17 killed 14% of the buds from undisturbed and 31% from cut-down plants of this ecotype.

The results presented above are based on the hardiness of lateral vegetative buds. Generally both the apex, when not abscised, and the lateral buds were slightly hardier than other tissues (pith, xylem, cortex). Visual observations of tissue browning after a 7-day regrowth period also indicated that hardening of the various stem tissues developed at slightly different rates. In samples collected in early October, pith was injured easier than cortex. Later, the situation was the opposite. These differences were quite small, but they were observed both in 1985 and 1986.

Table 4
COLD HARDINESS OF BUDS (LT_{50}) FROM VARIOUS ECOTYPES OF *S. PENTANDRA* GROWN OUT-OF-DOORS AS AFFECTED BY THE POSITION ON THE SHOOT

Position, cm from the top	LT_{50} (°C) for the buds: October 4, 1985, 70°	64°	60°	70° × 60°	October 23, 1985, 70°	64°	60°	70° × 60°
00-05	−13	−5	−4.5	−6.5	<−33	−22	−15	−24
05-10	−16	−8	−4.5	−8	<−33	−30	−18	−27
10-15	−17	−8	−4.5	−8	<−33	−30	−18	−25
15-20	—	—	—	—	—	−30	−19	−22
20-25	—	—	—	—	—	−28	−19	−24

Shoots used in these studies were about 1 to 1.5 m (1985) and 0.5 to 1 m long (1986). It could be expected that hardiness developed at different rates along the shoot, and this possibility was tested by taking series of samples from the top and toward the basis. As shown in Table 4, in the material from 1985 there was just a minor gradient of increasing hardiness toward the basis. The results from the experiments in 1986 showed, however, a more marked gradient. For example, combined results for two ecotypes (64°N latitude and the cross 70° × 60°N) gave the following percentages of injured buds in the samples collected on October 17 and frozen down to −25°C: 0 to 5 cm: 88%; 5 to 10 cm: 58%; 10 to 15 cm: 66%; 15 to 20 cm: 67%; 20 to 25 cm: 40%; 25 to 30 cm: 32%. The LT_{50} values were, respectively, −22.5, −24.3, −24.3, −23.3, −25.8, and −29.0°C ($LSD_{0.05}$ = 3.6). These gradients were rather similar for all the ecotypes studied and there were no significant interactions with the time of collection. The results in Figure 5 are therefore means of three or four topmost sections.

Apical growth cessation in seedlings of *S. pentandra*, induced by short photoperiod or by an improper light treatment, is followed by an increased cold hasrdiness even when the plants are not exposed to low temperature (Table 2). Short day treatment (12-hr photoperiod) which was continued over several weeks led to shedding of leaves even at 15°C and there was a significant increase of cold hardiness also at this temperature (Table 5). During an 8-week short day treatment cold hardiness of the buds increased from −8 to below −40°C at all temperature treatments used. The buds were hardier than stem tissues and after 8 weeks the cold hardiness of the pith was about −24°C. The results in Table 5 also indicate that an alternating temperature treatment (15/6°C) induced hardiness more effectively than constant temperature treatments. However, the positive interaction between short day treatment and low temperature for development of hardiness is very obvious (Tables 2, 5, and results presented by Kaurin[53]). Also, growing seedlings of the norther ecotype of *S. pentandra* obtain increased hardiness when exposed to low temperature (Table 2), but generally the ability for hardening is clearly negatively related to the visible elongation activity. In addition, there is the possibility that, for example, light treatments may affect cold hardiness even when there are no visually observable effects on elongation growth. The results in Table 2, confirming those presented earlier,[49] indicate that daily 12 hr blue light treatment may be able partly to prevent the cold hardening in *S. pentandra*. Also, results from an experiment with 1 hr end-of-day treatments combined with various photoperiods indicate a light effect on cold hardiness which is not related to the cessation of apical growth (Table 6).

Our results with the northern ecotype of *S. pentandra* are not so easy to interpret in terms of three phasic hardening processes.[38] The effect of photoperiod is rather marked and a high level of hardiness is obtained even without exposing the plants to freezing temperatures.

Table 5
DEVELOPMENT OF COLD HARDINESS IN SEEDLINGS OF *S. PENTANDRA* (ECOTYPE FROM 70°N LATITUDE) AT VARIOUS TEMPERATURE TREATMENTS UNDER 12 HR PHOTOPERIOD

	LT_{50} (°C) of the buds after n weeks			
Temperature	**0**	**2**	**4**	**8**
6	−8	−12	< −16 (16)	< −40 (16)
9	−8	−9	< −16 (14)	< −40 (8)
12	−8	−12	< −16 (11)	< −40 (4)
15	−8	−12	< −16 (13)	< −40 (4)
15/6[a]	−8	−13	< −16 (0)	< −40 (0)

Note: Seedlings were raised at 18°C and exposed to 12 hr photoperiod for induction of growth cessation before temperature treatments. After 8 weeks plants were completely defoliated at all temperature treatments. Figures in parenthesis give percentage of injured buds at −16°C and at −40°C.

[a] 12 hr/12 hr

Table 6
EFFECT OF 1 HR END-OF-DAY TREATMENTS WITH RED (R), FAR-RED (FR) AND BLUE (B) LIGHT ON BUD-SET AND COLD HARDINESS IN BUDS OF *S. PENTANDRA*

			LT_{50} (°C) for buds after	
Light treatment	**Photoperiod (hr)**	**Bud-set %**	**3 weeks at 18°C**	**3 weeks at 18°C + 3 weeks at 5°C**
1 hr R	21	100	−7	−13
1 hr FR	21	74	−6	−12
1 hr B	21	95	−7	−12
1 hr R	18	100	−12	−19
1 hr FR	18	100	−9	−22
1 hr B	18	100	−10	−20
				($LSD^{0.05}$:4.9)

Note: Ecotype from 70°N latitude.

However, the effect of freezing temperatures on hardening of *S. pentandra* has not be analyzed in any detail.

During the period of increasing cold hardiness in the fall, tissues of *S. pentandra* are very resistant to dehardening. Samples collected on October 27, 1986 and incubated at different temperature treatments (from 6 to 18°C) for 14 days did not show any damage in the stem tissue after freezing down to −20 to −30°C. Some of the buds started to grow during incubation and these buds were killed at −5°C. Nongrowing buds survived as well as stem tissue. Development of the bud dormancy in *S. pentandra* during the winter has not been investigated in detail, but the dormancy seems to be broken to a great extent during November. The capacity to obtain maximum hardiness is obviously not dependent on the stage of physiological dormancy. Also, the results with *S. polaris* from 78°N and 70°N latitude indicated that the bud dormancy was broken by a relatively short chilling treatment, and that the chilling requirement was shortest for the most northern ecotype.[51]

V. CONCLUDING REMARKS

Plant species growing in the Arctic and Subarctic are adapted to complete their annual growth cycle during a short and cold summer season with extreme long day conditions. In northern grass and woody species growth and dry matter production are strongly stimulated by long photoperiods. On the other hand, species like the herbaceous cloudberry and adult plants of woody species develop an endogenous annual growth rhythm which is only slightly affected by photoperiod.

Development of cold hardiness in northern grass species is primarily induced by low temperature, and investigations carried out so far do not indicate any light-dependent reactions which would be specific for the northern grass species. The same is true with cloudberry. These species, usually wintering under snow cover, do not develop a very high level of cold hardiness. Viviparous plantlets of arctic and subarctic grasses are, however, exceptions being capable of hardening down below −40°C. Cold hardiness in these structures should be investigated more thoroughly.

Cold hardiness in northern ecotypes of *S. pentandra*, as well as in other woody species, increases very rapidly during fall. This acclimation is clearly dependent on the preceeding cessation of elongation growth. This process is, at least in young plants, primarily regulated by photoperiod, although both temperature and spectral energy distribution can significantly modify the photoperiodic reactions. Both growth cessation and hardening are under genetic control, and crosses between northern and southern ecotypes show intermediate responses both with respect to growth cessation and cold hardiness.

Results presented in this paper clearly indicate that northern ecotypes of *S. pentandra* can obtain a high level of hardiness (below −40°C for the buds) at temperatures from 6° to 15°C when kept under short day conditions (12-hr photoperiod). Thus, they do not have an absolute requirement for low temperature for induction of a considerable level of hardiness.

Another point of interest is the effect of blue light. In the *Salix* ecotype from 70°N an increased level of blue light induced a partial growth cessation, but the cold hardiness did not increase as much as in other treatments showing a similar growth behavior. These results indicate an inhibition of acclimation in *Salix* by the blue light. With present knowledge it is difficult to see how this response might have any adaptive significance for plants growing under the northern conditions.

REFERENCES

1. **Kallio, P.,** The essence of biology in the North, *Nordia,* 18, 53, 1984.
2. **Hustich, I.,** Ecological concepts and biogeographical zonation in the North: the need for a generally accepted terminology, *Holarctic Ecol.,* 2, 208, 1979.
3. **Nilsen, J.,** Light climate in northern areas, in *Plant Production in the North,* Kaurin, Å., Junttila, O., and Nilsen, J., Eds., Norwegian University Press, Oslo, 1985,62.
4. **Larcher, W. and Bauer, H.,** Ecological significance of resistance to low temperature, in *Physiological Plant Ecology I, Responses to the Physical Environment,* Lange, O. L., Nobel, P. S., Osmond, C. B., and Ziegler, H., Eds., Springer-Verlag, Berlin, 1981, 403.

5. **Sakai, A. and Eiga, S.**, Physiological and ecological aspects of cold adaptation in boreal conifers, in *Plant Production in the North*, Kaurin, Å., Junttila, O., and Nilsen, J., Eds., Norwegian University Press, Oslo, 1985, 157.
6. **Heide, O. M.**, Physiological aspects of adaptation in plants with special references to high-latitude environments, in *Plant Production in the North*, Kaurin, Å., Junttila, O., and Nilsen, J., Eds., Norwegian University Press, Oslo, 1985, 1.
7. **Håbjørg, A.**, Effects of photoperiod and temperature on vegetative growth of different Norwegian ecotypes of *Poa pratensis, Meld. Norg. LandbrHøgsk.*, 55, 16, 1976.
8. **Endal, A., Vekst og Utvikling hos tre Arktisk-Alpine Gras;** *Deschampsia alpina* (L.) R. & S., *Poa alpina* L. *var. alpina* L. og *Festuca vivipara* (L.) Sm., Cand. real. thesis, University of Tromsø, Tromsø, Norway, 1982.
9. **Heide, O. M.**, Studies on flowering in *Poa pratensis* L. ecotypes and cultivars, *Meld. Norg. Landbr.Høgsk.*, 59, 14, 1980.
10. **Heide, O. M.**, Effects of photoperiod and temperature on growth and flowering in Norwegian and British timothy cultivars *(Phleum pratense* L.), *Acta Agric. Scand.*, 32, 241, 1982.
11. **Hay, R. K. M., and Heide, O. M.**, Specific photoperiodic stimulation of dry matter production in a high-latitude cultivar of *Poa pratensis, Physiol. Plant.*, 57, 135, 1983.
12. **Heide, O. M., Bush, M. G., and Evans, L. T.**, Interaction of photoperiod and gibberellin on growth and photosynthesis of high-latitude *Poa pratensis, Physiol. Plant.*, 63, 135, 1985.
13. **Eagles, C. F. and Østgård, O.**, Variation in growth and development in natural populations of *Dactylis glomerata* from Norway and Portugal. I. Growth analysis, *J. Appl. Ecol.*, 8, 367, 1971.
14. **Larsen, A. and Kaurin, Å.**, Unpublished data, 1986.
15. **Andersen, I. L. and Østgård, O.**, Virkning av Daglengde og Nattbelysning på Vekst, Stoffinnhold, Morfologi og Overvintring hos Noen Grasarter, *Forsk. For. Landbruket*, 31, 417, 1980.
16. **Håbjørg, A.**, Climatic control of floral differentiation and development in selected latitudinal and altitudinal ecotypes of *Poa pratensis* L., *Meld. Nor. Landbrukshoegsk.*, 57, 7, 1977.
17. **Lee, J. A. and Harmer, R.**, Vivipary, a reproductive strategy in response to environmental stress ?, *Oikos*, 35, 254, 1980.
18. **Junttila, O.**, Experimental control of flowering and vivipary in timothy *(Phleum pratense), Physiol. Plant.*, 63, 35, 1985.
19. **Andersen, I. L.**, Investigation on the wintering of meadow plants in Northern Norway, *Forsk. Fors. Landbruket*, 11, 635, 1960.
20. **Gudleifsson, B. E.**, Måling av Isdekketoleranse hos Gras i Laboratoriet, Nordiska Jordbruksforskares Forening, Seminar nr. 84, 171, 1986.
21. **Tronsmo, A. M.**, Induced resistance to biotic factors in grasses by frost hardening, in *Plant Production in the North*, Kaurin, Å., Junttila, O., and Nilsen, J., Eds., Norwegian University Press, Oslo, 1985, 127.
22. **Vestman, G.**, Urval for Resistens mot Vintringssvampar i Timotej, Nordiska Jordbruksforskares Forening, Seminar no. 84, 111, 1986.
23. **Hides, D. H.**, Winter hardiness in *Lolium multiflorum* Lam. III. Selection for improved cold tolerance and its effect on agronomic performance, *Grass Forage Sci.* 34, 119, 1979.
24. **Larsen, A.**, Freezing tolerance in grasses. Variation within populations and response to selection, *Meld. Norg. Landbr.Høgsk.*, 58, 42, 1979.
25. **Larden, A.**, Response to selection for freezing tolerance and associated effects on vegetative growth in *Dactylis glomerata*, in *Plant Production in the North*, Kaurin, Å., Junttila, O., and Nilsen, J., Eds., Norwegian University Press, Oslo, 1985, 134.
26. **Lorenzetti, F., Tyler, B. F., Cooper, J. P., and Breese, E. L.**, Cold tolerance and winter hardiness in *Lolium perenne*. I. Development of screening techniques for cold tolerance and survey of geographical variation, *J. Agric. Sci.*, 76, 199, 1971.
27. **Levitt, J.**, *Responses of Plants to Environmental Stresses*, Academic Press, New York, 1972.
28. **Larsen, A.**, Freezing tolerance in grasses. Methods for testing in controlled environments, *Meld. Norg. Landbr.Høgsk.*, 57, 23, 1978.
29. **Lawrence, T., Cooper, J. P., and Breese, E. L.**, Cold tolerance and winter hardiness in *Lolium perenne*. II. Influence of light and temperature during growth and hardening, *J. Agric. Sci.*, 80, 341, 1973.
30. **Kaurin, Å.**, Effects of light quality on frost hardning in *Poa alpina*, in *Plant Production in the North*, Kaurin, Å., Junttila, O., and Nilsen J., Eds., Norwegian University Press, Oslo, 1985, 116.
31. **Kaurin, Å.**, Unpublished data, 1986.
32. **Junttila, O., Nilsen, J., and Rapp, K.**, Research on cloudberry in Norway, *Metsäntutkimuslaitoksen Tiedonantoja*, 90, 23, 1983.
33. **Bottengård, E.**, Virkning av Temperatur og Iordfuktighet på Vekst og Utvikling hos Molte *(Rubus chamaemorus* L.), Cand. real. thesis, University of Tromsø, Norway, 1980.
34. **Kaurin, Å., Stushnoff, C., and Junttila, O.**, Vegetative growth and frost hardiness of cloudberry *(Rubus chamaemorus)* as affected by temperature and photoperiod., *Physiol. Plant.*, 55, 76, 1982.

35. **Rapp, K. and Stushnoff, C.,** Artifical freezing of *Rubus chamaemorus* L. for estimation of genetic components of cold hardiness, *Meld. Norg. Landbr.Høgsk.*, 58, 15, 1979.
36. **Kaurin, Å., Junttila, O., and Hansen J.,** Seasonal changes in frost hardiness in cloudberry *(Rubus chamaemorus)* in relation to carbohydrate content with special reference to sucrose, *Physiol. Plant.*, 52, 310, 1981.
37. **Stushnoff, C. and Junttila, O.,** Seasonal development of cold stress resistance in several plant species at a coastal and a continental location in North Norway, *Polar Biol.*, 5, 129, 1986.
38. **Weiser, C. J.,** Cold resistance and acclimation in woody plants, *HortScience*, 5, 403, 1970.
39. **Christersson, L.,** The influence of photoperiod and temperature on the development of frost hardiness in seedlings of *Pinus sylvestris* and *Picea abies*, *Physiol. Plant.*, 44, 288, 1978.
40. **Paulsen, G. M.,** Effects of photoperiod and temperature on cold hardening in winter wheat, *Crop. Sci.*, 8, 29, 1968.
41. **Vaartaja, O.,** Photoperiodic ecotypes of trees, *Can. J. Bot.*, 32, 392, 1954.
42. **Dormling, I., Gustafsson, A., and von Wettstein, D.,** The experimental control of the life cycle in *Picea abies* (L.) Karst. Some basic experiments on the vegetative cycle, *Silvae Genet.*, 17, 44, 1968.
43. **Heide, O. M.,** Growth and dormancy in Norway spruce ecotypes *(Picea abies)*. I. Interaction of photoperiod and temperature, *Physiol. Plant.*, 30, 1, 1974.
44. **Håbjørg, A.,** Photoperiodic ecotypes in Scandinavian trees and shrubs, *Meld. Norg. Landbr.Høgsk.*, 57, 33, 1978.
45. **Junttila, O.,** Effect of photoperiod and temperature on apical growth cessation in two ecotypes of *Salix* and *Betula*, *Physiol. Plant.*, 48, 347, 1980.
46. **Junttila, O.,** The cessation of apical growth in latitudinal ecotypes and ecotype crosses of *Salix pentandra* L., *J. Exp. Bot.*, 33, 1021, 1982.
47. **Koski, V.,** Adaptation of trees to the variation in the length of the growing season, in *Plant Production in the North*, Kaurin, Å., Junttila, O., and Nilsen, J., Eds., Norwegian University Press, Oslo, 1985, 267.
48. **Håbjørg, A.,** Effects of light quality, light intensity and night temperature in growth and development of three latitudinal populations of *Betula pubescens* Ehrh., *Meld. Norg. Landbr.Høgsk.*, 51, 2, 1972.
49. **Junttila, O. and Kaurin, Å.,** Climatic control of apical growth cessation in latitudinal ecotypes of *Salix pentandre* L., in *Plant Production in the North*, Kaurin, Å., Junttila, O., and Nilsen, J., Eds., Norwegian University Press, Oslo, 1985, 83.
50. **Junttila, O.,** Apical growth cessation and shoot tip abscission in *Salix*, *Physiol. Plant.*, 38, 278, 1976.
51. **Paus, E., Nilsen, J., and Junttila, O.,** Bud dormancy and vegetative growth in *Salix polaris* as affected by temperature and photoperiod, *Polar Biol.*, 6, 91, 1986.
52. **Paus, E.,** Regulering av Hvile og Vekst hos *S. polaris* WG., Cand. sci. thesis, University of Tromsø, Tromsø, Norway, 1984.
53. **Kaurin, Å.,** Freezing Mechanisms and Climatic Effects on Frost Hardening in Some Woody and Herbaceous Plants, Ph.D. thesis. University of Tromsø, Tromsø, Norway, 1985.

Chapter 8

LOW TEMPERATURE INJURY IN WINTER CEREALS

M. K. Pomeroy and C. J. Andrews

TABLE OF CONTENTS

I. Introduction 108

II. Cold Acclimation and Freezing Damage 108

III. Low Temperature Flooding 110

IV. Ice Encasement 113

V. Conclusions 119

References 120

I. INTRODUCTION

Winter cereals are grown over a wide geographical range in temperate regions of the world and consequently are subjected to a diverse array of low temperature stresses. In the more northerly latitudes of the winter wheat growing areas where snow accumulation is not abundant, overwintering damage may result from direct freezing injury. However, in the major winter wheat growing areas, winter temperatures are usually moderate or snow cover is sufficient throughout the winter to protect the crop from freezing damage. In northeastern North America, autumn flooding, and ice encasement due to mid-winter thaws and rain is common, and the occurrence of these conditions usually results in overwintering damage. Fungal pathogens (snow molds) also are responsible for significant winterkill in areas where snow accumulates and temperatures hover near the freezing point for an extended period in spring.

Field and controlled environment experiments have delineated the role of various natural and artificially induced low temperature stresses on survival of winter cereals. In controlled freeze tests, cold acclimated winter wheats can withstand temperatures in the range of -12 to -22°C, whereas the most freezing tolerant winter rye cultivars can survive brief exposure to temperature as low as -30°C.[1,2] An increase in the duration of freezing stress significantly decreases survival of winter cereals,[3,4] probably due to disruption of key metabolic processes in the roots.[5,6] Flooding at low temperature gradually reduces viability of winter cereals and is associated with a gradual accumulation of ethanol and CO_2.[7] In contrast, total ice encasement rapidly damages winter wheat and is accompanied by a rapid rise in levels of these key metabolites.[8,9] More recent experiments[10] have shown that exposure of winter wheat to various combinations of low temperature flooding and freezing stress induces much greater damage than the single stresses.

Although the physiological responses of winter cereals to various low temperature stresses has been thoroughly documented, the underlying mechanisms of adaptation and injury are not well understood. Considerable evidence has accumulated suggesting that the site of freezing injury in plants may be the plasma membrane, and that cold acclimation involves changes in structural and functional properties of these membranes.[11-13] The mechanism(s) of damage due to low temperature flooding and ice encasement is less well understood, but it must be different from freezing damage since injury may occur near 0°C without ice formation in the cells or tissue dehydration. It has been suggested that damage to winter cereals under these conditions results from anoxia-induced accumulation of cellular metabolites, such as ethanol and CO_2, to toxic levels.[7,9,14]

This chapter summarizes the effects of freezing, low temperature flooding, and ice encasement stresses on winter cereals under field as well as controlled environment conditions. In addition, evidence for current hypotheses on mechanisms of low temperature injury are examined and avenues for future research in this area are discussed.

II. COLD ACCLIMATION AND FREEZING DAMAGE

Winter cereal species have the capacity to develop varying degrees of tolerance to freezing stress after several weeks of cool autumn temperatures under natural field conditions, or by exposing young seedlings to temperatures in the range of 0 to 5°C for a few weeks under controlled environment conditions. Cold acclimated winter rye cultivars attain significantly greater freezing tolerance than either winter wheat or barley and, hence, usually exhibit greater winter survival. Winter wheats are generally more resistant to freezing damage than winter barleys, but even the most cold resistant wheat cultivars cannot withstand the low temperature stresses prevalent in many of the northern temperate regions of the world. For this reason, the areas in which winter wheat can successfully be grown is limited, and significant winterkill occurs in marginal growing areas. However, expansion of winter wheat production in many areas of the world, including Canada, is desirable since winter cultivars are generally more productive than spring-sown wheats and they provide an excellent alternative for crop rotations.

Although it has been recognized for hundreds of years that plants are adversely affected by low temperature and thousands of papers have been published on this subject, we still have an incomplete understanding of the mechanisms involved in cold acclimation and freezing damage. Since this subject has been thoroughly reviewed over the years,[12,13,15-19] we will limit our discussion in this chapter to some of the more recent developments in this area. It is now generally accepted that the plasma membrane is a primary site of injury due to freezing stress. Numerous reports suggest that one form of freeze-thaw injury results from irreversible loss of membrane material.[13] Williams and co-workers[20,21] suggested that an important element of freezing injury in winter wheat is the loss of membrane material, particularly lipid, from cell membranes, and that cold acclimation may involve changes in the cell that facilitate return of deleted materials to the membranes. These observations have been extended in recent years by investigations on protoplasts and isolated cells of winter rye. Steponkus and co-workers investigated in detail the osmotic behavior of protopolasts subjected to freeze-thaw treatments using a variety of techniques, including conventional light microscopy, transmission and scanning electron microscopy, freeze fracture, and cryomicroscopy coupled to a video display.[22-24] It was demonstrated that the surface area of the plasma membrane of protoplasts from nonacclimated winter rye seedlings is not conserved during freeze-thaw induced osmotic contraction and expansion. During contraction, endocytotic vesiculation occurs and membrane material is transferred from the plasma membrane to membrane reservoirs in the cytoplasm. However, this material is not readily reincorporated into the plasma membrane during protoplast thawing and expansion and lysis usually occurs before the protoplast is able to return to its original volume. In protoplasts from cold acclimated seedlings, freeze-induced dehydration and contraction results in the formation of polyp-like exocytotic extrusions from the plasma membrane, but these extrusions are readily drawn back into the plasma membrane during thaw-induced expansion, and the protoplasts do not lyse. Injury to acclimated protoplasts was attributed to the loss of osmotic responsiveness of the protoplasts rather than membrane lysis. The exact mechanism whereby low temperature acclimation alters membrane properties is not known, but may be due to changes in composition of the plasma membrane.

The investigations of Singh and co-workers[11,25] and of Pearce and Willison[26] have provided further evidence of the involvement of the plasma membrane in cold acclimation and freezing injury in winter cereals. Transmission electron microscopy of frozen fixed, isolated nonhardy rye cells showed that regions of plant membranes, such as the plasma membrane, can fuse to form multilayered vesicles. This reduces the amount of membrane material available for cell expansion during rehydration, often resulting in cell lysis on thawing. The nature of changes in membrane structure which results in the formation of multilamellar vesicles in nonhardened cells is not fully understood, but it has been suggested[26] that intramembrane particle free regions formed on the plasma membrane during freezing are sites for the irreversible formation of multilamellar vesicles. In cold hardened rye cells frozen fixed at $-10°C$, folding and invagination of the plasmalemma were observed, but no multibilayered membranes could be discerned. Also, the volume to surface area ratio of the plasmalemma of hardened rye cells is uncoupled during plasmolysis in balanced salt solutions by the formation of cytoplasmic strands between the cell wall and the protoplast. During deplasmolysis, and presumably during rehydration following a nonlethal freeze, this membrane material is re-incorporated into the expanding protoplast without damaging the cell.

The observations discussed above provide convincing evidence of the involvement of changes in the properties of the plasma membrane in the development of freezing tolerance in winter cereals. However, it is not yet clear whether these changes are the primary event in this phenomenon, or if they are a manifestation of other changes in cellular metabolism. For example, recent investigations on structural and functional properties of winter rye chloroplasts[27,28] indicated that growth at cold-hardening temperatures results in an increased capacity for photosynthetic electron transport. It was concluded that this change in function was associated with alterations in thylakoid membrane protein complexes. Enhanced photosynthetic electron transport during growth at low temperature in light may be crucial to the development of maximum levels of freezing tolerance as it can provide energy necessary for the cold acclimation process. However, this does not appear to be a primary event in the overall process, since winter cereal seedlings also develop relatively high levels of freezing tolerance during growth at low temperature in the dark[29] by utilizing carbohydrates stored in the endosperm.

The chemical modifications which occur in cellular membranes during cold acclimation and which minimize the deleterious effects of freezing to sublethal temperatures have not yet been fully elucidated. Williams et al.[21] observed that the lipids from cold hardy Kharkov winter wheat have properties very different from those of unhardened Kharkov, or other less hardy cultivars. The authors suggested that a solubilizing agent, probably a cytoplasmic lipid-exchange protein, may be involved in reabsorption of deleted membrane material during rehydration following a freezing event. Recent investigations of Uemura and Yoshida[30] on winter rye revealed distinct differences between polypeptide components, sterol composition, phospholipid-to-protein ratio, and sterol-to-phospholipid ratio from plasma membranes of nonacclimated and acclimated seedlings. It was suggested that changes in protein composition and phospholipid enrichment may be more important than sterol changes in the development of cold hardiness. Cold acclimation induced changes in polypeptide composition of soluble proteins of winter wheat and rye also have been reported.[31] It is clear from these observations that the chemical composition of cell membranes is in a dynamic state, and that cold acclimation induces significant structural alterations. However, we do not yet understand the manner in which these alterations are related to the observed changes in response of the plasma membrane to a freeze-thaw event. The preponderance of evidence does suggest that the key to understanding the molecular mechanism(s) of cold acclimation and freezing injury in winter cereals is to further elucidate the changes in structural and functional properties of the plasma membrane associated with these phenomena.

Recent studies with two different plant species[32,33] have demonstrated the production of specific cold-induced messenger RNAs, which translate new polypeptides during cold hardening, but such work has not yet been reported for winter wheat. However, Sarhan and his co-workers[34] have determined that cold tolerant wheats synthesize more total RNA at low temperatures than cold susceptible wheats. The RNA levels are regulated by increases in, and changing ratios of RNA polymerases I and II involved in DNA-dependent elaboration of ribosomes, and mRNA, respectively. More work is required to identify, isolate, and characterize the genetic material directing cold hardening in cereals before the techniques of genetic manipulation can be utilized to improve the freezing tolerance of winter cereals.

III. LOW TEMPERATURE FLOODING

Overwintering crops are susceptible to damage by excessive water at any time during their seedling growth in fall, winter or spring (Figure 1). Flooding in fall and spring in most central and northern latitudes inundates plants in cold water. Although the duration of exposure depends on a number of environmental and edaphic factors, the most significant one is soil drainage. Fall flooding of long duration has the potential of forming ice sheets, while the duration of spring flooding from rain or snow melt may be dependent upon the thawing of surface drainage systems. The effects of flooding on crop plants are wide ranging

A

B

FIGURE 1. Photographs of a winter wheat plot in Ottawa showing that the low areas which were flooded in spring (A) suffered severe overwintering damage (B).

and the subject has recently been reviewed for plants at growing temperature.[35] Only studies which have a direct relation to overwintering problems will be discussed here.

Flooding of grasses at low temperature is generally less damaging than it is at high temperature.[36] At high temperature, enzyme activity and enzyme induction in response to declining oxygen levels in the medium is rapid[37] and metabolic pathways leading to injurious conditions are activated. At low temperatures, these processes are markedly slower resulting in gradual damage over a period of several days, but may occur over a period of months as has been shown for some forage grasses.[38] Solubility of oxygen in water is higher at low temperatures than at high temperatures, which serves to retard the advance of anoxia. Winter crops may be exposed to the adverse effects of flooding for many weeks at a time, and the cumulative responses of the plants are significant to their survival.

Total immersion of cold acclimated cereal plants in water at 2°C for 2 weeks markedly reduces cold hardiness.[39] The reductions are of a similar magnitude in winter barley and two varieties of winter wheat, but the barley which has lower initial hardiness is reduced to very low cold resistance by the flooding stress. Increase in flooding depth leads to a greater effect on cold hardiness. Hardiness is only slightly reduced unless the plant crown is submerged, and maximum reduction occurs with total submergence of crown and leaf. Flooding at the soil surface during low temperature growth reduces the capacity of seedlings of less hardy cultivars to acclimate,[40] but acclimation of the very hardy Norstar wheat is only slightly affected (Figure 2). Moisture contents of plant crowns decrease during low temperature growth even in flooded conditions, but the levels attained are higher than those of nonflooded plants (Figure 2). Moisture contents closely parallel the cold hardiness of the tissue either during acclimation, or during flooding treatments of fully acclimated plants. In either case, the capacity of affected plants to withstand future freezing stress is markedly diminished. Flooding at low temperature, however, increases the tolerance of wheat plants to subsequent ice encasement at −1°C,[39] although not at −5°C[41] (see following section). Low temperature growth does not acclimate plants to low temperature flood stress[42] as it does to freezing and icing stresses. Studies on fifteen cultivars grown at low temperature have shown that no correlation exists between flooding tolerance and freezing tolerance, while correlations between flooding and icing tolerances are significant.[43]

Anaerobic products accumulate in winter cereal crowns during flooding at 2°C[7] and at 5°C.[44] Ethanol, CO_2 and lactic acid accumulate under controlled environment conditions and in flooded field conditions,[45] while total carbohydrate content of plant crowns declines.[44] The anaerobic products accumulate as survival of plants decreases, but no causative relationships have yet been established. Survival of winter barley declines more rapidly than winter wheat, and the barley produces significantly more CO_2 and lactic acid, but not ethanol, than the wheat. Jackson et al.[46] reported that during flooding at median temperatures (20°C), ethanol concentrations are far too low to account for death of plants by ethanol toxicity. Both alanine[47] and asparagine[48] accumulate during low temperature flooding, which may be an adaptive response to stress. In contrast, acetaldehyde[49] has been shown to accumulate in cereal crowns during flooding at 15°C and to have marked toxicity to wheat tissues. Peroxidation of membrane lipids during the immediate post-flood period has been implicated in plant injury at 20°C.[50]

Ultrastructure of the apicial cells of winter cereals is altered by flooding at low temperature, with winter barley showing more rapid deterioration of structure than winter wheat.[7] This includes the rupture of membranes, breakdown of cellular organelles, and the elaboration of cisternal endoplasmic reticulum. Such effects are a manifestation of the greater sensitivity of barley to various products of anaerobiosis at low temperature[51] leading to the more rapid kill of barley during flooding stress.

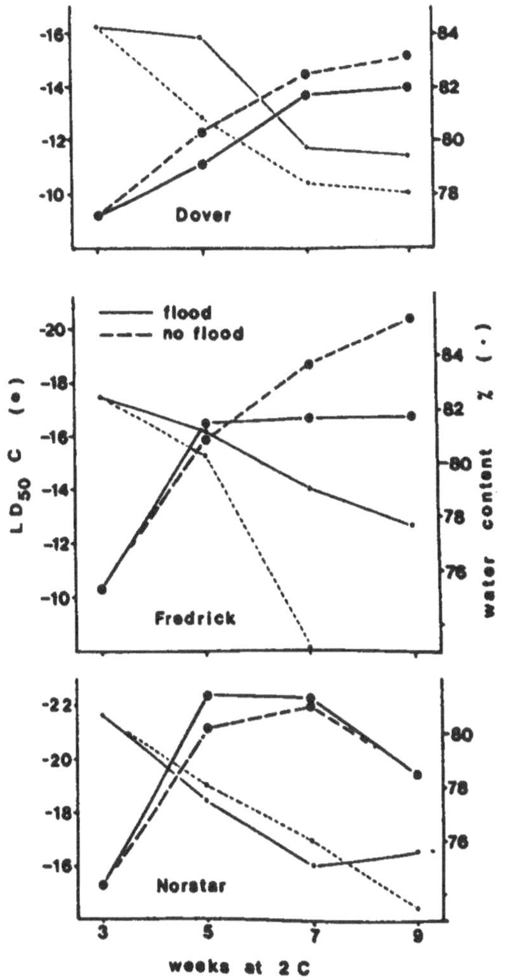

FIGURE 2. The increase in cold hardiness (LD_{50}°C) and decrease in moisture content of 3 winter cereals during low temperature growth with and without flooding to the soil level.

IV. ICE ENCASEMENT

Covering of low-growing plants by ice during winter has been recognized as a damaging condition for many years.[52] This may occur from fall flooding or more frequently due to a winter thaw in snow-prone areas. Melted water or rain percolates through the snow, builds up at the soil surface, and refreezes with the return of low winter temperatures (Figure 3). Ice may be more or less solid depending upon the prevailing snow structure and the totality of the thaw process. Penetration of solid ice into the soil depends on soil moisture content at the time of freeze-up, and the amount of thaw water and the duration of the thaw.

Winter crops in regions such as most northern maritime areas, western North America, northern Europe, and the western parts of the Soviet Union with relatively heavy winter precipitation and temperatures which fluctuate around the freezing point are vulnerable to ice encasement, and in some areas the growth of winter cereals is completely prevented by icing. Although well-drained sandy soils are less likely to form damaging ice layers, drainage capacity of such soils is severely limited after freeze-up.

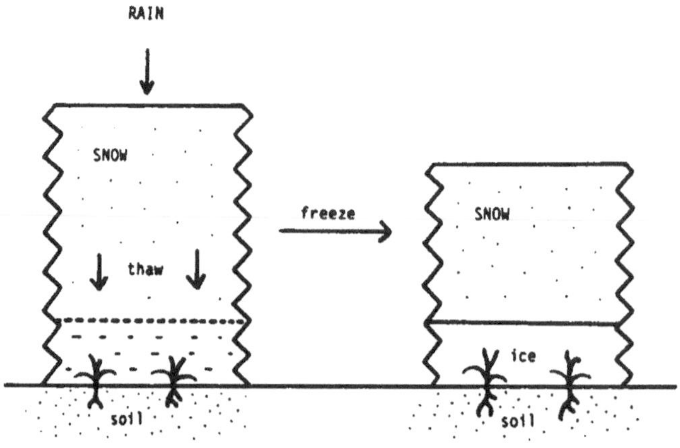

FIGURE 3. The formation of ice layers under snow during midwinter thaw.

FIGURE 4. Overwintering damage due to ice encasement in low areas of a field of winter wheat.

Overall, ice encasement is severely injurious to crop plants (Figure 4). Early studies described damage to forage grasses,[38,53] legumes,[54,55] and cereals[56] and demonstrated some cultivar variation in resistance to the stress. Much of this variation is caused by time of onset of the stress (the ice forming event) during the winter. In many areas this controls the duration of the stress because the ice usually does not dissipate until advent of the spring thaw. One of the major difficulties in determing the extent and severity of ice encasement damage in field conditions is that the ice formation occurs at the soil surface, normally under a snow layer and is thus obscured from direct observation. The ice layer may not even be apparent in spring, as the ice can deteriorate rapidly during snow melt, and its integrity be lost during the alternate freezing and thawing which is a feature of the northern spring. Partly because of this, the term "smothering" has frequently been used[57] to describe damage to crops occurring as a result of restricted gaseous exchange through deep snow layers with or without ice. Discrimination between the two causes of damage is difficult because of problems observing the sequence of events at the soil surface beneath a snow cover.

Widespread damage to overwintering crops associated with heavy ice encasement in parts of eastern Canada after the winter of 1972, prompted us to initiate investigation on icing damage to winter cereals using conditions prevailing in the affected areas. Temperature of plant crowns under ice and insulating snow in Ontario were found rountinely to be around −1°C.[58] However, if insulating snow cover was removed from winter wheat plots, the temperature of plants under the ice dropped rapidly resulting in greater damage.[58,59] At −1°C in total ice encasement in controlled enviroments, kill of fully hardened winter wheat seedlings occurred in 1 to 3 weeks. However, if the plants were only partially ice encased with leaves protruding through the ice,[8,59] or roots were in soil with low moisture content,[60,61] survival was not reduced for several weeks. A similar response was observed in alfalfa, where stubble from older plants allowed aeration to living plants beneath an ice layer.[62]

Resistance to ice encasement increases with duration of low temperature growth, and is thus an inducible response. Winter cereals differ in their capacity to develop ice tolerance and the response is correlated, but not closely, with the development of cold hardiness ($r = 0.69$, $n = 30$). Partial ice encasement has little effect on levels of subsequent cold hardiness, but total ice encasement reduces resistance to later cold stress markedly in both controlled environment[8] and in field conditions.[45] In practice, this reduces the ability of cereal seedlings to withstand further freezing stress, even if the ice layer is removed during winter by a major thaw.

Early studies on alfalfa indicated that the cause of death of plants in ice encasement may largely be due to accumulation of up to 24% CO_2.[54,63] Reports in the Russian literature described accumulations of ethanol in cereal seedlings under ice and in "smothering" conditions but there was divergence of opinion over the significance of such accumulations in relation to damage to plants. Rakitina demonstrated that ice was nearly impermeable to both oxygen and CO_2,[64] and reported depletions of O_2 and major accumulations of CO_2 (up to 40%) in the extractable gas from encased plants.[14] It was later shown that such levels of CO_2 associated with low oxygen in a gas mixture at below freezing temperatures reduced survival, regrowth, and subsequent cold hardiness of hardened plants.[65] The major effect of the ice encasement stress, or the equivalent gas regime was a limitation on the resistance of the cells of the crown and leaves to frost desiccation by increasing cell water permeability.[66] However, the effect of changing physiological response with differential cultivar survival was not determined.

Ethanol has been shown to accumulate rapidly in a number of winter cereal cultivars in ice at −1°C[66] (Figure 5) and at −2°C,[44] but comparison of times at which 50% of the cultivar populations were killed, showed that there was a higher level of ethanol in the cultivars with higher tolerance to stress.[67] Furthermore, at the 50% kill point due to ice exposure at −1°C, there was considerably less endogenous ethanol than was present in the tissues at the 50% kill point due to exogenous ethanol, indicating that ethanol cannot be the sole source of damage during ice encasement.[67] Carbon dioxide accumulates in approximately equimolar amounts as ethanol (Figure 5), but lactate accumulates in 1 day to a low stable level.[9] It is noteworthy that lactate accumulates to much higher levels in flooded conditions in the field[45] and in controlled environment[7] than it does in ice encasement. Lactate does not accumulate to a sufficient level in ice to account for plant death, but CO_2 accumulates to a higher level at the 50% kill point due to ice than at the 50% kill point due to exogenous CO_2, indicating that CO_2 accumulation can account for the death of plants in ice encasement. The effects of CO_2 with and without ethanol on survival, permeability to amino acids and electrolytes, and on the ultrastructure of crown meristem cells indicate that both ethanol and

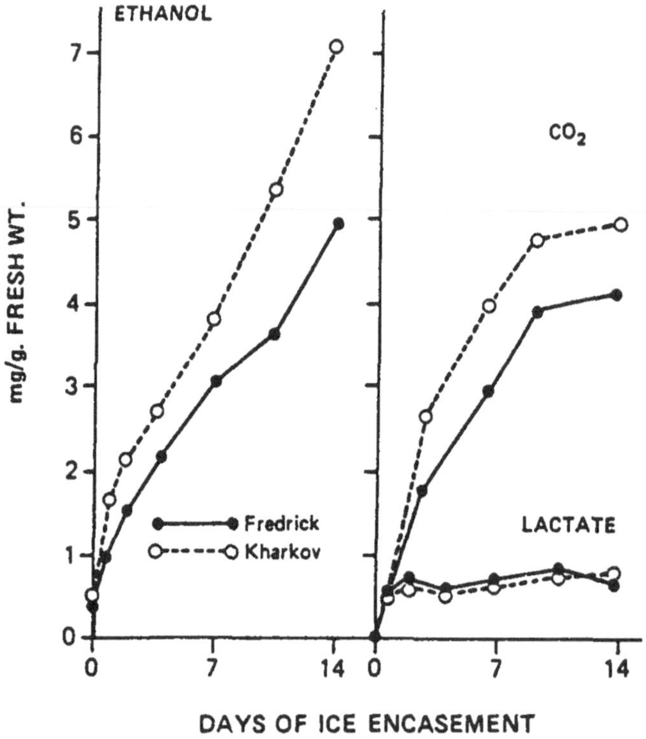

FIGURE 5. The accumulation of ethanol, CO_2 and lactic acid in crown tissue of cold hardened Fredrick and Kharkov winter wheats during total ice encasement at − 1°C.

CO_2 contribute to plant death in ice encasement.[9] Using ^{14}C-label in an anaerobic system at 5°C Hallam and Mckersie[68] showed that the ethanol and CO_2 pools are further metabolized and the label accumulates predominantly in an asparagine fraction of winter wheat crowns. Labeling of the asparagine fraction was correlated with the ice tolerance of nine wheat cultivars.[48]

Plants under the anaerobic stress of ice accelerate their utilization of carbohydrate in the crown meristem. Cultivars exhibiting greater tolerance to stress tend to utilize a greater proportion of carbohydrates.[44] As only 20 to 25% of total carbohydrate is utilized in a 1 to 2 week period of total icing, death of plants is not directly associated with a deficiency of carbohydrate reserves.[44,60] Together with decreases in total carbohydrate and ethanol soluble carbohydrate there are increases in reducing sugars during ice encasement[44,60] which include increases in a fructose component.[69]

A period of ice encasement which severely reduces survival of winter wheat plants causes little disruption of ultrastructure of cells in the apical meristem. In contrast, low temperature freezing (− 10°, − 16°C) sufficient to incur similar survival reductions causes major damage to cellular organization.[70,71] Similarly, using tetrazolium staining, Tanino and McKersie[72] showed no immediate localized damage to cell function after ice, but a steady decline in the post-ice period. The damaging effects of freezing stress on celluar function were confined to the vascular transition zone of the lower crown.

Mitochondrial activity and integrity have been shown to decline slowly in plants in ice at − 1°C, but at a stage when 50% of the plants are killed, respiratory properties are little impaired, and no major structural alterations occur until 3 weeks of icing.[29] The resistance of mitochondria to freezing stress has also been demonstrated by isolation of fully active mitochondria from freeze killed wheat tissue.[73]

In field conditions, ice encasement rarely occurs as a single stress, and is usually one of a complex of stresses impinging upon the plants in a winter season. Analysis of these interactions is not only important in understanding the timing and progress of overwintering damage, but can also provide clues to understanding the mechanism of ice encasement injury. When a period of low temperature flooding was followed by ice encasement at −5°C, simulating possible conditions in the winter wheat growing areas of the USSR, an increase in plant kill was observed as compared to that with ice encasement alone.[41] However, in our experiments if 1 to 4 weeks of low temperature flooding was followed by ice encasement at −1°C, an increase in survival in several varieties of winter wheat was observed but not in commercial barleys, despite major decreases in cold hardiness of both cereals.[39] This is an adaptive response of the wheat plant, which by exposure to the mild anaerobic stress of low temperature flooding or nitrogen exposure, increases its tolerance to the severe anaerobic stress of ice encasement. Greater sensitivity of barely to low temperature flooding,[51] and to ethanol[8] and other products of anaerobiosis eliminates the response in this cereal.

The increased survival during ice encasement after flooding is accompanied by higher levels of alcohol dehydrogenase activity and a higher rate of ethanol production, which indicates that previously flooded plants enter ice encasement with higher glycolytic potential than nonflooded plants. Substrate level (glycolytic) phosphorylation is the major source of energy for cellular maintenance during ice encasement with its attendant low oxygen levels. However, by the end of the flooding period, total adenylates and adenylate energy charge (AEC) are lower in flooded than in nonflooded plants. It is only after a short period of ice encasement that the nonflooded plants undergo large decreases in total adenylates and AEC, to levels lower than in the previously flooded plants which retain their levels of these parameters.[74]

Sub-freezing temperatures in the range of −6 to −10°C reduce tolerance of wheat plants to ice encasement, but do not reverse the promotive effect of a previous flooding treatment.[10] Flooding increases the resistance of plant crowns to freezing-induced leakage of amino acids after a freezing stress and also increases the uptake of ^{86}Rb, an analog of potassium, into root and shoot tissues.[10]

Light exposure during ice encasement increases survival of seedlings of a number of winter cereals in comparison with icing in dark. Light decreases ethanol production by plants in ice, increases CO_2 production but reduces carbohydrate utilization.[75] Slightly higher levels of total adenylates and AEC have been observed in iced plants in light than in dark.[74] Oxygen is produced by photosynthetic activity at ice encasement temperature, thus precluding total anaerobiosis.[75] Oxygen levels of 3 to 4% after several weeks of ice encasement in short day conditions were reported by Rakitina.[14]

An isolated cell system has been employed to investigate further the effects of ice stress on cell permeability and the characteristics of the plasma membrane.[76] The cells were found to have similar survial characteristics to the primary leaf tissue from which they were isolated, but were more tolerant to ice stress than the intact plants from which both were derived (Figure 6).[77] Some cells remained alive after 42 days of ice, indicating that the cell system is a potentially powerful selection technique. Declining survival during ice exposure at −1°C was closely associated with leakage of amino acids through the plasma membrane, but an early and major effect was a decrease in the uptake of ^{86}Rb (Figure 7). This indicates that inhibition or injury of membrane transport could be the initial lesion due to ice encasement, leading to eventual death of the cells. Ion transport is inhibited by exogenous ethanol but the inhibition only occurs at greater than physiological concentrations. Physical characteristics of the membranes, as determined by electron spin resonance spectroscopy, are not influenced by ice encasement or by low levels of ethanol.

Isolated cells of wheat are partially protected against damage during ice encasement by calcium (10m*M*) which also reverses the inhibitory effect of ice exposure on the ^{86}Rb uptake

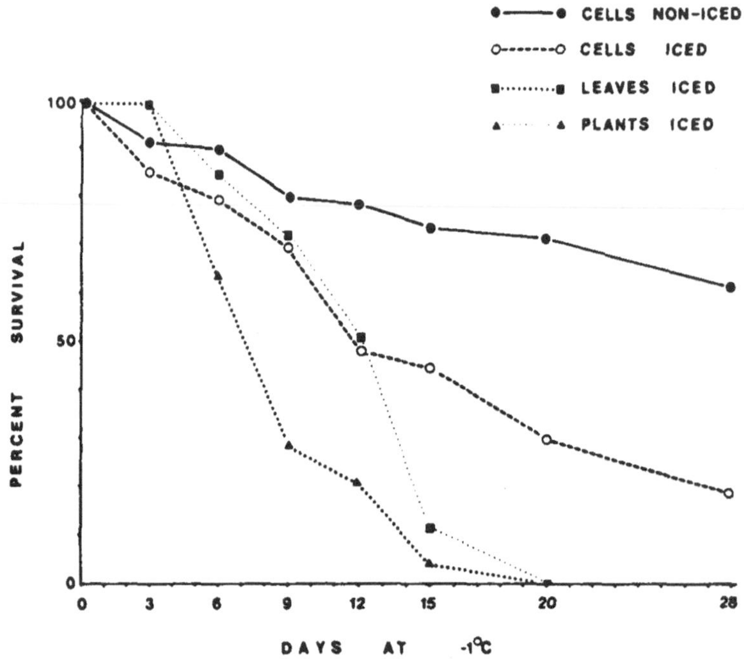

FIGURE 6. Survival of isolated cells, detached leaves and intact plants of cold hardened Kharkov winter wheat during ice encasement at − 1°C, in comparison with noniced cells at − 1°C.

system (Figure 7).[78] The effect is ion specific, with a slight duplication of the effect observed only with strontium.[79] Calcium has little effect on protecting cells from low temperature freezing stress, indicating that the modes of injury induced by the two stresses are of a different nature,[74] although both have their effect on the plasma membrane.

Damage to the ion transport system by ice encasement could result from a reduction in energy charge below the level required for active ion uptake, or from inhibition of the plasma membrane ATPase(s) involved in this process. It has been estimated that at least 20% of cellular ATP is utilized by membrane ATPases.[80] In isolated wheat cells, ATP does not decrease rapidly enough during ice encasement to account for the loss in ion uptake after 1 day of ice exposure. Rather, it decreases gradually paralleling the decline in cell survival in ice.[81] AEC does not decline in ice encasement, presumably indicating that the small demands on ATP supply for synthetic processes in surviving ice-stressed cells are adequately met by the ATP production of fermentation.

If one assumes that substrate deficiency is not the cause of membrane transport inhibition, blockage of enzyme activity by inhibition or denaturation of a rapidly accumulating anaerobic product remains a possibility. Ethanol has been implicated,[67] but it does not have inhibitory activity at low concentrations.[76] Carbon dioxide has a major effect on plant survival at low temperature and is a component of cellular acidosis described as contributing to cell death due to anaerobiosis.[82,83] Plasma membrane enriched fractions isolated from tissues ice encased for increasing periods indeed show declining levels of ATPase activity while the activity fron noniced tissue at − 1°C remains high. The activity of these membranes is, however, severely inhibited by carbonate or biocarbonate ions in solution, but much less so by ethanol and lactic acid.[84]

Membrane transport capacity of isolated cells is reduced more rapidly than cell survival during icing, and more closely approximates the decline in survival of intact plants and crowns. It is apparent that in cells while the lesion in membrane function exists after short

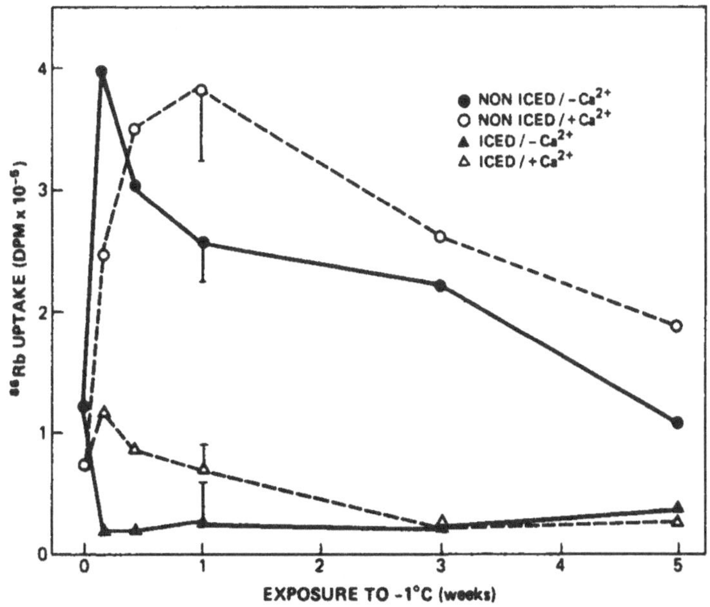

FIGURE 7. Uptake of ^{86}Rb by isolated cells of cold hardened Kharkov winter wheat immediately after eposure to -1°C with or without ice encasement, and in the presence or absence of 10 m*M* calcium.

periods of ice exposure, it is not immediately effective in the absence of external ions, and in the absence of significant communication between the cell and its frozen environment. Intact plants damage more rapidly, presumably due to perturbation of the complex organization of the plant crown, which requires intimate cellular communication for its integrity and continuing function.

V. CONCLUSIONS

Overwintering damage is a major limitation to the production of winter cereals in northern latitutes. Although major strides have been made in the development of agronomically suitable cultivars with improved winter tolerance, it appears that further improvements through conventional plant breeding will be difficult. Application of rapidly developing technologies of genetic manipulation appear to offer the possibility of transferring genes for cold hardiness into less hardy winter cereals. Before this can be achieved, however, we must gain a thorough understanding of the physiology and molecular biology of acclimation to low temperature stresses and cold injury. The available evidence suggests that damage to winter cereals due to freezing, low temperature flooding, and ice encasement involves the plasma membrane, but the mechanisms involved are still not clearly understood, and considerably more research is required in these areas before the development of superior cultivars by genetic manipulation can be realized.

REFERENCES

1. **Andrews, C. J., Pomeroy, M. K., and Seaman, W. L.,** The response of fall-sown cereals to winter stresses in eastern Ontario, *Can. J. Plant Sci.*, 66, 25, 1986.
2. **Fowler, D. B., Dvorak, J., and Gusta, L. V.,** Comparative cold hardiness of several *Triticum* species and *Secale cereale* L., *Crop Sci.*, 17, 941, 1977.
3. **Pomeroy, M. K., Andrews, C. J., and Fedak, G.,** Cold hardening and dehardening responses in winter wheat and winter barley, *Can. J. Plant Sci.*, 55, 529, 1975.
4. **Gusta, L. V. and Fowler, D. B.,** Factors affecting the cold survival of winter cereals, *Can. J. Plant Sci.* 57, 213, 1977.
5. **Chen, T. H.-H., Gusta, L. V., and Fowler, D. B.,** Freezing injury and root development in winter cereals, *Plant Physiol.*, 73, 773, 1983.
6. **Pomeroy, M. K., Andrews, C. J., Stanley, K. P., and Gao, J-Y.,** Physiological and metabolic responses of winter wheat to prolonged freezing stress, *Plant Physiol.*, 78, 207, 1985.
7. **Pomeroy, M. K. and Andrews, C. J.,** Metabolic and ultrastructural changes associated with flooding at low temperature in winter wheat and barley, *Plant Physiol.*, 64, 635, 1979.
8. **Andrews, C. J. and Pomeroy, M. K.,** Survival and cold hardiness of winter wheats during partial and total ice immersion, *Crop Sci.*, 15, 561, 1975.
9. **Andrews, C. J. and Pomeroy, M. K.,** Toxicity of anaerobic metabolites accumulating in winter wheat seedlings during ice encasement, *Plant Physiol.*, 64, 120, 1979.
10. **Gao, J.-Y., Andrews, C. J., and Pomeroy, M. K.,** Interactions among flooding, freezing, and ice encasement in winter wheat, *Plant Physiol.*, 72, 303, 1983.
11. **Johnson-Flanagan, A. M. and Singh, J.,** Membrane deletion during plasmolysis in hardened and non-hardened plant cells, *Plant Cell Environ.*, 9, 299, 1986.
12. **Steponkus, P. L.,** Cold hardiness and freezing injury in agronomic crops, in *Advances in Agronomy*, Vol. 30, Brady, N. C., Ed., Academic Press, New York, 1978, 51.
13. **Steponkus, P. L.,** Role of the plasma membrane in freezing injury and cold acclimation, *Annu. Rev. Plant Physiol.*, 35, 543, 1984.
14. **Rakitina, Z. G.,** Effect of an ice crust on gas composition of the internal atmosphere of winter wheat, *Sov. Plant Physiol.*, 17, 907, 1970.
15. **Levitt, J.,** *The Hardiness of Plants,* Academic Press, New York, 1956.
16. **Levitt, J.,** *Responses of Plants to Environmental Stresses,* Academic Press, New York, 1972.
17. **Levitt, J.,** *Responses of Plants to Environmental Stresses,* Vol. 1, Chilling, freezing, and high temperature stresses, Academic Press, New York, 1980.
18. **Olien, C. R.,** Freezing stresses and survival, *Annu. Rev. Plant Physiol.*, 18, 387, 1967.
19. **Burke, M. J., Gusta, L. V., Quamme, H. A., Weiser, C. J., and Li, P. H.,** Freezing and injury in plants, *Annu. Rev. Plant Physiol.*, 27, 507, 1976.
20. **Williams, R. J. and Hope, H. J.,** The relationship between cell injury and osmotic volume reduction. III. Freezing injury and frost resistance in winter wheat, *Cryobiology,* 18, 133, 1981.
21. **Williams, R. J., Willemot, C., and Hope, H. J.,** The relationship between cell injury and osmotic volume reduction. IV. The behaviour of hardy wheat membrane lipids in monolayer, *Cryobiology,* 18, 146, 1981.
22. **Steponkus, P. L., Dowgert, M. F., Ferguson, R. L., and Levin, R. L.,** Cryomicroscopy of isolated plant protoplasts, *Cryobiology,* 21, 209, 1984.
23. **Dowgert, M. F. and Steponkus, P. L.,** Behaviour of the plasma membrane of isolated protoplasts during a freeze-thaw cycle, *Plant Physiol.*, 75, 1139, 1984.
24. **Gordon-Kamm, W. J. and Steponkus, P. L.,** The behaviour of the plasma membrane following osmotic contraction of isolated protoplasts: implications in freezing injury, *Protoplasma,* 123, 83, 1984.
25. **Singh, J. and Miller, R. W.,** Biophysical and ultrastructural studies of membrane alterations in plant cells during extracellular freezing: molecular mechanism of membrane injury, in *Cryopreservation of Plant Cells and Organs,* Kartha, K. K., Ed., CRC Press, Boca Raton, Fla., 1985, 61.
26. **Pearce, R. S. and Willison, J. H. M.,** Wheat tissues freeze-etched during exposure to extracellular freezing: distribution of ice, *Planta,* 163, 295, 1985.
27. **Huner, N. P. A.,** Acclimation of winter rye to cold-hardening temperatures results in an increased capacity for photosynthetic electron transport, *Can. J. Bot.*, 63, 506, 1985.
28. **Griffith, M. Huner, N. P. A., and Hayden, D, C.,** Low temperature development of winter rye leaves alters the detergent solubilization of thylakoid membranes, *Plant Physiol.*, 81, 471, 1986
29. **Andrews, C. J. and Pomeroy, M. K.,** Mitochondrial activity and ethanol accumulation in ice-encased winter cereal seedlings, *Plant Physiol.*, 59, 1174, 1977.
30. **Uemura, M. and Yoshida, S.,** Involvement of plasma membrane alterations in cold acclimation of winter rye seedlings (*Secale cereale* L. cv. Puma), *Plant Physiol.*, 75, 818, 1984.
31. **Cloutier, Y.,** Changes in the electrophoretic patterns of the soluble proteins of winter wheat and rye following cold acclimation and desiccation stress, *Plant Physiol.*, 71, 400, 1983.

32. **Guy, C. L., Niemi, K. J., and Brambl, R.**, Altered gene expression during cold acclimation of spinach, *Proc. Natl. Acad. Sci. U. S. A.* 82, 3673, 1985.
33. **Johnson-Flanagan, A. and Singh, J.**, Induction of freezing tolerance in suspension cultures of *Brassica napus* by ABA at 25°C: studies on protein and RNA changes, in *Plant Cold Hardiness*, Li, P. H., Ed., Plant Biology Series, A. R. Liss, New York, 1987.
34. **Sarhan, F. and Chevrier, N.**, Regulation of RNA synthesis by DNA-dependent RNA polymerases and RNases during cold acclimation in winter and spring wheat, *Plant Physiol.*, 78, 250, 1985.
35. **Kozlowski, T. T., Ed.**, *Flooding and Plant Growth*, Academic Press, New York, 1984.
36. **Beard, J. B. and Martin, D. P.**, Influence of water temperture on submersion tolerance of four grasses, *Agron. J.*, 62, 257, 1970.
37. **Hake, S., Kelly, P. M., Taylor, W. C., and Freeling, M.**, Coordinate induction of alcohol dehydrogenase 1, aldolase and other anaerobic RNAs in Maize, *J. Biol. Chem.*, 260, 5050, 1985.
38. **Beard, J. B.**, Effects of ice, snow and water covers on kentucky bluegrass, annual bluegrass and creeping bentgrass, *Crop Sci.*, 4, 638, 1964.
39. **Andrews, C. J. and Pomeroy, M. K.**, The effect of flooding pretreatment on cold hardiness and survival of winter cereals in ice encasement. *Can. J. Plant Sci.*, 61, 507, 1981.
40. **Andrews, C. J. and Pomeroy, M. K.**, Effects of flooding on cold hardiness of winter cereals, *Plant Physiol.*, 63S, 88, 1979.
41. **Rakitina, Z. G.**, Effects of a surrounding ice crust on winter wheat plants as a function of their flooding before freezing. *Sov. Plant Physiol.*, 254, 317, 1977.
42. **McKersie, B. D.**, Acclimation of winter cereals for cold icing and flooding stresses. *Annu. Rept. Univ. of Guelph Crop Sci. Dept.*, 1983, 105.
43. **McKersie, B. D.**, Types of winter stresses — do winter wheat cultivars respond differently?, in *New Frontiers in Winter Wheat Production*, Fowler, D. B., Gusta, L. V., Slinkard, A. E., and Hobin, B. A., Eds., University of Saskatchewan, Saskatoon, 1983, 26.
44. **McKersie B. D., McDermott, B. M., Hunt, L. A., and Poysa, V.**, Changes in carbohydrate levels during ice encasement and flooding of winter cereals, *Can. J. Bot.*, 60, 1822, 1982.
45. **Pomeroy, M. K. and Andrews, C. J.**, Metabolic and ultrastructural changes in winter wheat during ice encasement under field conditions, *Plant Physiol.*, 61, 806, 1978.
46. **Jackson, M., Berman, H., and Goodenough, A.**, An examination of the importance of ethanol in causing injury to flooded plants, *Plant Cell Environ.*, 5, 163, 1982.
47. **Ostaplyuk, E. D. and Beletskaja, E. K.**, Adaptation and resistance of winter crops to anaerobiosis, *Int. Symp. Plant Anaerobiosis*, 1985, 20 (in Russian).
48. **McKersie, B. D.**, Personal communication.
49. **Beletskaja, E. K.**, Changes in metabolism of winter crops during their adaptation to flooding, *Sov. Plant Physiol.*, 24, 750, 1977.
50. **Hunter, M. I. S., Hetherington, A. S., and Crawford, R. M. M.**, Lipid peroxidation — a factor in anoxia intolerance in *Iris* species, *Phytochem.*, 22, 1145, 1983.
51. **Andrews, C. J. and Pomeroy, M. K.**, The influence of flooding pretreatment on metabolic changes in winter cereal seedlings during ice encasement, *Can. J. Bot.*, 61, 142, 1983.
52. **Wright, S. G.**, Relation of low temperature to the growth of wheat, *Agric. Sci.*, 4, 337, 1980.
53. **Sjoseth, H.**, Studies on frost hardening in plants, *Acta Agric. Scand.*, 14, 178, 1964.
54. **Sprague, V. G. and Graber, L. F.**, Physiological factors operating in ice sheet injury to alfalfa, *Plant Physiol.*, 15, 661, 1940.
55. **Smith, D.**, The survival of winter-hardened legumes encased in ice, *Agron. J.*, 44, 469, 1952.
56. **Salmon, S. C.**, Why cereals winterkill, *J. Am. Soc. Agron.*, 9, 353, 1917.
57. **Lamb, C. A.**, Physiology, in *Wheat and Wheat Improvement*, Quisenberry, K. S. and Reitz, L. P., Eds., American Society of Agronomy, Madison, Wis., 1967, 181.
58. **Andrews, C. J. and Pomeroy, M. K.**, Changes in survival of winter cereals due to ice cover and other simulated winter conditions, *Can. J. Plant Sci.*, 57, 1141, 1977.
59. **Rakitina, Z. G.**, Influence of temperature on plant survival under conditions of ice encasement, *Fiziol. Rast.*, 29, 1053, 1982.
60. **Pomeroy, M. K. and Andrews, C. J.**, Flooding and ice encasement damage to winter wheat, in *New Frontiers in Winter Wheat Production.*, Fowler, D. B., Gusta, L. V., Slinkard, A. E., and Hobin, B. A., Eds., University of Saskatchewan, Saskatoon, 1983, 39.
61. **Bolduc, R.**, personal communication.
62. **Freyman, S.**, Role of stubble in the survival of certain ice covered forages, *Agron. J.*, 61, 105, 1969.
63. **Sprague, M. A. and Graber, L. F.**, Ice sheet injury to alfalfa, *J. Am. Soc. Agron.*, 35, 881, 1943.
64. **Rakitina, Z. G.**, The permeability of ice for O_2 and CO_2 in connection with a study of the reasons for winter cereal mortality under the ice crust, *Sov. Plant Physiol.*, 12, 795, 1965.
65. **Rakitina, Z. G.**, Effect of the gaseous composition of the atmosphere on the frost resistance of winter wheat plants, *Sov. Plant Physiol.*, 14, 278, 1967.

66. **Samgsin, G. A., Rakitina, Z. G., and Livshin, A. Z.,** Influence of an adverse gaseous regime and ice-crust on the water retaining capacity of the leaves and tillering nodes of hardened winter wheat, *Dok. Bot. Sci.*, 209, 60, 1973.
67. **Andrews, C. J.,** Accumulation of ethanol in ice-encased winter cereals, *Crop Sci.*, 17, 157, 1977.
68. **Hallam, P. and McKersie, B. D.,** Metabolism of ^{14}C-sucrose in roots of winter cereals during anaerobiosis, *Can. Soc. Plant Physiol. Proc.*, 20, 1983.
69. **Andrews, C. J.,** unpublished results.
70. **Pomeroy, M. K. and Andrews, C. J.,** Ultrastructural changes in shoot apex cells of winter wheat seedlings during ice encasement, *Can. J. Bot.*, 56, 786, 1977.
71. **Singh, J. and Andrews, C. J.,** Comparisons of ultrastructural changes during extracellular freezing at − 10°C and ice encasement at − 1°C in winter wheat crown cells by the method of freeze fixation, *Cryoletters*, 2, 117, 1981.
72. **Tanino, K. K. and McKersie, B. D.,** Injury within the crown of winter wheat seedlings after freezing and icing stress, *Can. J. Bot.*, 63, 432, 1985.
73. **Singh, J., de la Roche, I. A., and Siminovitch, D.,** Relative insensitivity of mitochondria in hardened and nonhardened rye coleoptile cells to freezing *in situ*, *Plant Physiol.*, 60, 713, 1977.
74. **Andrews, C. J. and Pomeroy, M. K.,** in preparation.
75. **Andrews, C. J.,** The increase in survival of winter cereal seedlings due to light exposure during ice encasement, *Can. J. Bot.*, 66, 409, 1988.
76. **Pomeroy, M. K., Pihakaski, S. J., and Andrews, C. J.,** Membrane properties of isolated winter wheat cells in relation to icing stress, *Plant Physiol.*, 72, 535, 1983.
77. **Andrews, C. J. and Pomeroy, M. K.,** Ice encasement tolerance of isolated cells from cold hardened winter wheat, *Cryoletters*, 4, 33, 1983.
78. **Pomeroy, M. K. and Andrews, C. J.,** Effect of low temperature and calcium on survival and membrane properties of isolated winter wheat cells, *Plant Physiol.*, 78, 484, 1985.
79. **Pomeroy, M. K.,** unpublished results.
80. **Penning de Vries, F. W. T.,** The cost of maintenance processes in plant cells, *Ann. Bot.*, 39, 77, 1975.
81. **Pomeroy, M. K. and Andrews, C. J.,** Changes in adenine nucleotides and energy charge in isolated winter wheat cells during low temperature stress, *Plant Physiol.*, 81, 361, 1986.
82. **Roberts, J. K. M., Callis, J., Jardetsky, O., Walbot, V., and Freeling, M.,** Cyloplasmic acidosis as a determinant of flooding intolerance in plants, *Proc. Natl. Acad. Sci. U.S.A.*, 81, 6029, 1984.
83. **Roberts, J. K. M., Andrade, F. H., and Anderson, I. C.,** Further evidence that cytoplasmic acidosis is a determinant of flooding intolerance in plants, *Plant Physiol.*, 77, 492, 1985.
84. **Pomeroy, M. K. and Andrews, C. J.,** unpublished results.

Chapter 9

PLASMA MEMBRANE ATPase AS A SITE OF FUNCTIONAL ALTERATION DURING COLD ACCLIMATION AND FREEZING INJURY

S. Iswari and Jiwan P. Palta

TABLE OF CONTENTS

I. Introduction 124

II. The Hypothesis 124

III. The Basis of the Hypothesis 125
A. Post-Thaw Recovery 125

IV. Investigations on the Activity of Plasma Membrane ATPase After Cold Acclimation and Freezing Injury 126
A. Changes in Plasma Membrane ATPase After Freezing Injury and Cold Acclimation in Potato Species 127
1. Standardization of a Procedure to Purify Plasma Membranes 127
2. Effect of Freeze Thaw and Cold Acclimation on Plasma Membrane ATPase Activity 132

V. Conclusions 135

Acknowledgment 135

References 135

I. INTRODUCTION

Plasma membrane as a site of freezing injury was recognized as early as 1912.[1] However, the nature of the injury to the membrane at a more fundamental level has been investigated only recently.[2-8] This perhaps has been due to the assumption that freezing injury results in a complete loss of membrane semipermeability or membrane rupture. This notion has persisted even in recent literature.[9] Contrary to this view, Palta and co-workers[2-4,10-12] have demonstrated that incipient freezing injury results in an alteration in membrane transport properties. Evidence was presented implicating alteration in the active transport systems (presumably ATPases) located in the cell membrane in the initial stages of freezing injury.[2,11,12] At that time a K^+-stimulated Mg^{2+} requiring ATPase was known to exist in the plant plasma membrane.[13] Several studies since then have implicated alteration in the properties of plasma membrane ATPase after freezing injury and cold acclimation.[7,14-19] Recently it has been suggested that the plasma membrane ATPases are in fact H^+-ATPases which are involved in ion transport across the membrane.[20] Using the recent methods for purifying the plant plasma membrane we have obtained pure plasma membrane preparations and investigated the alterations in the properties of plasma membrane ATPase in potato species due to freeze-thaw injury and cold acclimation. We present here a brief account of the background information and summarize our preliminary results supporting the hypothesis that freeze-thaw injury results in alteration of the active transport properties of the cell membrane.

In nature, during a frost episode, herbaceous plants experience slow cooling (a rate of about 1°C/hr). In spite of the well recognized fact that freezing injury may vary depending upon the cooling rates[8], many recent studies have used very fast cooling rates (10 to over 1000°C/hr).[21-24] Unfortunately, from some of these studies several hypotheses of freezing injury have been advanced.[5] Clearly the relevance of these hypotheses to explain freezing injury in herbaceous plants under natural conditions is, at best, extremely limited. In our studies we simulated more closely the natural freezing process by using slow cooling rates. The relevance of our results to natural freeze stress should be viewed from this perspective.

II. THE HYPOTHESIS

Palta and co-workers[2-4,10-12] have proposed that a sublethal freeze stress (using slow cooling rates) results in an alteration of the transport properties of the cell membranes. This is an early manifestation of the injury resulting from a slow freeze-thaw stress. These alterations result from functional or structural changes in the membrane associated H^+-ATPase as a result of the stress, although how these changes are brought about is not clearly understood. Palta and Li[3] suggested that these may result from: (1) direct denaturation of membrane transport proteins or (2) changes in lipid-protein interaction. In recent years it has become clear that the lipid bilayer is a major factor in determining the function of the integral membrane proteins.[25] Various features of the membrane bilayer lipids that can drastically alter the activity of transport proteins include lipid head groups, lipid acyl chain length, saturation/unsaturation, lipid backbone, and bilayer fluidity.[25] Alteration in the functions of H^+-ATPases following a sublethal freeze-thaw injury could be brought about by such changes in the membrane lipids associated with these proteins. Recently an alteration of the lipid component of the membrane following cold acclimation and freezing injury has been reported.[26-28] An alteration in the active transport properties could be brought about by changes in the membrane lipids closely associated with H^+-ATPases. However, conformational alterations of membrane proteins could also explain this injury. Recently Arora and Palta have implicated a role for Ca^{2+} in freezing injury.[29] Intracellular calcium is thought to be a "second messenger" which can influence many metabolic functions through its effect on calmodulin.[30] Changes in the functions of H^+-ATPases could be brought about by changes in the intracellular concentration of Ca^{2+}. Future work at the membrane and cellular level is necessary to elucidate the molecular sequence of events during freezing injury.

III. THE BASIS OF THE HYPOTHESIS

The hypothesis outlined above was proposed by Palta and co-workers and is based on several studies conducted with onion bulb and potato leaf tissue.[2-4,10-12] The following important results were shown from these studies.

An enhanced efflux of ions following freezing injury was not accompanied by membrane rupture. In spite of enhanced net efflux of ions, (primarily K^+) all the cells could be plasmolyzed and these cells could be stained with a fluorescent vital stain (uranin) and they exhibited cytoplasmic streaming. Electron microscopic examination of freeze injured (about 50% of total ion leaked out) onion bulb cells and potato leaf palisade cells showed no ultrastructural abnormalities compared to unfrozen control. The cell membranes appeared normal and intact. All these observations demonstrated that membrane rupture or a complete loss of semipermeability is not a prerequisite for ion leakage following freezing injury.

In spite of enhanced ion efflux following freezing injury, there was no significant change in the permeability of the cell membrane to water and nonelectrolytes. However, there was an increase in passive permeability to K^+ depending on the degree of injury. Using freeze injured yet alive cells, the membrane permeability to water and nonelectrolytes such as urea and methyl urea was measured. Although there was a dramatic increase in K^+ efflux in these cells, no change in the passive permeability to water or nonelectrolytes was found. Water and nonelectrolytes are known to be transported passively via the lipid portion of the membrane.[31] Therefore, these results suggested that transport properties of membrane lipids following a freeze-thaw stress remained unaltered.

While there was no significant change in the passive permeability to water and nonelectrolytes, significant increase in the passive permeability to K^+ was found. Since K^+ transport across the membrane is thought to be associated with intrinsic membrane proteins, the increased K^+ permeability after freezing suggested an alteration of the functions of the intrinsic membrane proteins involved in K^+ transport. There is ample evidence in the literature that K^+ transport across the cell membrane is coupled to H^+-ATPases.[20] A change in K^+ transport properties following injury therefore could be explained in terms of freeze-thaw induced perturbation of the functions of the H^+-ATPases.

A. Post-Thaw Recovery

Study of the freeze injured onion cells during the post-thaw period provided another independent basis for the hypothesis outlined above. Depending on the degree of injury, a complete recovery of the injury accompanied by a complete disappearance of water infiltration of the tissue was observed. Visually no difference could be detected between the freeze injured-recovered tissue and unfrozen control. During the recovery a simultaneous decrease in K^+ efflux was found. Freeze injured cabbage[32] and potato[4] showed similar recovery and disappearence of water infiltration. Furthermore, as in onion tissue, in potato leaves a decrease in K^+ efflux was found during recovery.[4]

The recovery must be an active process. Leakage of ions due to freezing injury results in a decrease in concentration gradient between the vacuole and extracellular water resulting in a flaccid appearance of the tissue. During recovery an opposite process must take place, i.e., the ions have to be transported back into the cell which then leads to uptake of extracellular water (regaining the turgor). Since ion influx (during recovery) is against the concentration gradient it must be an active process. The recovery from injury thus requires recovery of the functions of an active transport system. The injured cells may have the ability to repair whatever damage had occurred. It may be renaturation of partially denatured proteins or fresh synthesis of the proteins and assembly into the membrane or regaining of the normal lipid-protein interactions.

IV. INVESTIGATIONS ON THE ACTIVITY OF PLASMA MEMBRANE ATPase AFTER COLD ACCLIMATION AND FREEZING INJURY

Subsequent to the proposal of the hypothesis by Palta and co-workers,[2-4,10-12] a few groups have independently investigated the plasma membrane ATPase in relation to freezing injury and cold acclimation. The following account gives a critical summary of the information available and also summarizes the results of our work in tuber-bearing *Solanum* species.

Jian et al.[14] have studied the plasma membrane and cell organelle ATPase activities in relation to cold acclimation and freezing injury in winter wheat cells using a cytochemical method. They observed that during freezing injury the plasma membrane ATPase activity decreased before the organelle ATPase activities were affected. During cold acclimation they observed the opposite effect, i.e., the plasma membrane ATPase activity increased in parallel with frost tolerance.

Direct measurement of plasma membrane ATPase activity in relation to cold acclimation and freezing injury has been made using purified plasma membrane preparations from Pinus[7,16,17] and Jerusalem artichoke tubers.[18,19] The purification was based on the method of partitioning between aqueous two-phase polymer systems developed by Albertsson.[33] Hellergren et al.[16] prepared the plasma membranes by partitioning the 1000 to 30,000 x *g* pellet in 6.3% w/w each of polyethylene glycol (PEG) and dextran containing 5 m*M* Na-phosphate and 0.25*M* sucrose adjusted to pH 7.8. The plasma membranes which were recovered from the PEG-rich upper phase were characterized by the presence of K^+, Mg^{2+}-ATPase activity and silicotungstic acid staining for electron microscopy (marker for plasma membrane), cyt.c oxidase (marker for mitochondria) and antimycin A insensitive NADH-cyt.c reductase (marker for endoplasmic reticulum). The specific activity of the K^+, Mg^{2+}-ATPase in the upper phase membranes was ten times higher than in the lower phase. The extent of contamination by other marker enzymes cannot be assessed by their report, because there is no information on the total protein content of the fractions to enable a calculation of the total activity. They observed a similar pattern of changes in ATPase activity during acclimation as Jian et al.[14] observed in winter wheat, i.e., the specific activity of the ATPase in purified plasma membrane preparations increased during the end of the acclimation period and decreased during deacclimation.[16] After freezing to lethal temperature (LT_{50} of -7°C) the specific activity of the ATPase decreased to zero.[7] However, it is not clear from their report how the total protein content and total ATPase activity recovered in the plasma membrane fraction was affected. This information is necessary to understand more clearly the significance of the changes in ATPase activity after cold acclimation and freezing injury.

Ishikawa and Yoshida[18] have studied plasma membrane ATPase in tubers of *Helianthus tuberosus*. Plasma membranes were purified by aqueous two-phase partitioning. Plasma membranes were prepared by partitioning the 14,000 to 156,000 x *g* pellet in 5.6% each of PEG and dextran with 30 m*M* NaCl, 0.25*M* sucrose and 10 m*M* K-phosphate pH 7.8. The characterization of the plasma membrane was done by testing for ATPase activity, naphthyl-phthalamic acid binding, and PTA-CrO_3 staining (markers for plasma membrane), IDPase (marker for golgi), cyt.c oxidase (marker for mitochondria), antimycin A insensitive NADH-cyt.c reductase (marker for endoplasmic reticulum), and acid phosphatase. The plasma membranes were assessed to be of high purity. They observed an increase in specific activity of the ATPase during cold acclimation.[18] In a subsequent study Uemura and Yoshida[19] found that while the specific activity did not change greatly during freezing stress the total plasma membrane ATPase activity decreased considerably even at nonlethal freezing temperatures

(-5°C) and further deteriorated at a more severe freezing stress (-10°C). However, they found that the ATPase activity, recovered in the plasma membrane fraction prepared by sucrose density gradient centrifugation of microsomes, did not change significantly after lethal freezing. Hence they suspect that the changes in the plasma membrane ATPase activity observed after freezing (in membrane preparations isolated by the phase partitioning method) could be due to a change in surface properties of the plasma membrane. Such a change in surface properties could result in a changed partitioning behavior of the membranes. It is not clear from their report how the total microsomes obtained from a particular amount of tissue was affected by freezing, and how much of the microsomes was loaded on the sucrose density gradient. Furthermore, while they observed 75 to 80% inhibition by vanadate (a specific inhibitor of plasma membrane ATPase) of the plasma membrane ATPase in the membranes prepared by phase partitioning, there was only 35 to 40% inhibition by vanadate in the plasma membrane obtained by sucrose density gradient centrifugation. This suggests that sucrose density gradient method did not give them as pure a membrane preparation. Therefore, though there is a possibility that changes in plasma membrane ATPase activity observed after freezing could be due to a change in the surface properties of plasma membrane which results in a changed partitioning behavior, their results do not provide unequivocal evidence to the fact that it is indeed happening. Uemura and Yoshida[19] also observed an increased inhibition by DCCD of ATPase activity in frozen samples compared to the control which indicates that there may be a conformational change of ATPase or an increase in permeability of DCCD into the membrane.

The work of Pomeroy et al.[15] indicates quite clearly that damage to ion transport system could be one of the earliest manifestations of icing stress. They studied ^{86}Rb uptake (a measure of K^+ transport) in isolated winter wheat cells following icing and exposure to ethanol. They observed a decline in ^{86}Rb uptake before a decline in cell viability and oxygen consumption by the cells.

A. Changes in Plasma Membrane ATPase After Freezing Injury and Cold Acclimation in Potato Species

1. Standardization of a Procedure to Purify Plasma Membranes

Potato species provide a good system to study cold acclimation and freezing injury because of the availability of different species with significant differences in cold acclimating ability and freeze stress resistance.[34,35] The following account summarizes the establishment of a technique to prepare plasma membranes from the leaf tissue of two *Solanum* species — *S. tuberosum* and *S. commersonii*. Some of the preliminary results on the activity of plasma membrane ATPase in isolated plasma membranes in the two species after cold acclimation and freezing injury are also summarized.

A prerequisite for any investigation on plasma membrane is the establishment of a method for purifying the plasma membrane and validating its purity. The criteria we chose to test for the purity were from Serrano.[36] The ATPase activity should be insensitive to KNO_3, azide, and molybdate which are specific inhibitors of vacuolar ATPase, mitochondrial ATPase, and phosphatases respectively, and the activity should be inhibited by vanadate. Other markers we chose to test were cyt.c oxidase (for mitochondria), NADH-cyt.c reductase (for endoplasmic reticulum), and chlorophyll content (for chloroplasts).

So far, a majority of the isolations of the plasma membrane from plants have been performed from root tissue. There are very few reports of succesful isolations of plasma membranes from leaves. This is probably due to the difficulty faced in preparing plasma membranes free of chloroplast contamination, which come in an array of size and density once the cell is broken during isolation. Kjellbom and Larsson[37] have prepared chlorophyll-free plasma membranes from spinach and barley leaves. Their procedure involves partitioning the 10,000 to 50,000 × *g* pellet serially five times in a system of 6.2% (w/w) each of PEG

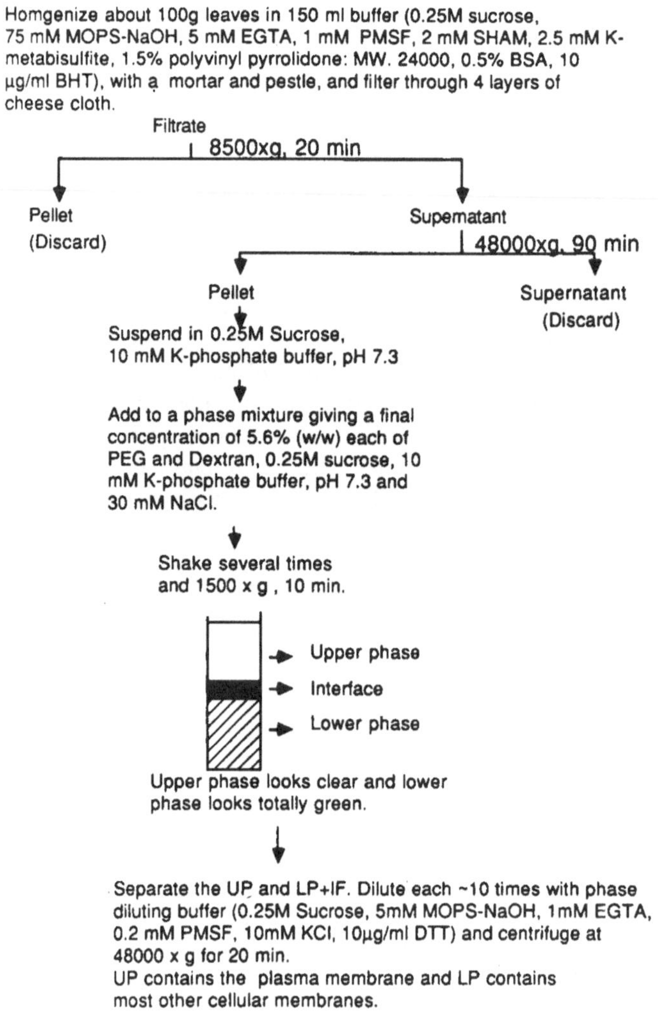

FIGURE 1. Preparation of plasma membrane from *Solanum* leaves.

and Dextran with 0.33 *M* sucrose, 3 m*M* KCl, and 5m*M* K-phosphate pH 7.8. When we used their procedure with *S. commersonii* leaves the results were discouraging. The upper phase membranes obtained after one step of partitioning were enriched in plasma membrane ATPase activity (72% of that in microsomes), but there was also a very heavy contamination by chloroplast membranes (82% of chlorophyll in microsomes). Therefore, this method was discontinued and we explored the use of various concentrations of different salts like NaCl, LiCl, and Li_2SO_4 to reduce the contaminations. Out of the three salts tested, 10m*M* NaCl was better as judged by the presence of vanadate sensitive ATPase activity (48% of total). Increasing the concentration of NaCl to 20m*M* further reduced all these contaminants and at 30 m*M* NaCl presence of the contaminants was negligible. We found similar results with *S. tuberosum* and *S. commersonii.* A simple flow-chart of the procedure used for plasma membrane preparation is shown in Figure 1. The technique has been repeated several times since, with similar results. The results of characterization of the membranes are shown in Tables 1 and 2. In both species the upper phase (UP) membranes obtained in presence of 30 m*M* NaCl were rich in plasma membrane ATPase activity and free from contamination

Table 1
EFFECT OF NaCl ON THE PARTITIONING OF *S. TUBEROSUM* PLASMA MEMBRANE IN A TWO-PHASE POLYMER SYSTEM

		ATPase activity			Cyt.c-oxidase		Cyt.c-reductase		Chlorophyll[d]		ATPase activity	
Fraction	Protein (mg)	SA[b]	Fold SA	% Activity	SA[c]	% Activity	SA[c]	% Activity	(mg)	(%)	SA[b] without LPC	Activation with LPC (fold SA[b])
Microsomes	37	0.13	1	100	0.031	100	0.039	100	1.00	100	0.086	1.6
10 m*M* NaCl												
UP	7.1	0.37	2.8	55	0.014	8.3	0.053	27	0.28	28	0.11	3.3
LP	7.0	0.20	1.5	29	0.064	41	0.088	44	0.58	58	0.10	2.0
20 m*M* NaCl												
UP	4.7	0.40	3.0	39	0.0069	2.1	0.052	17	0.10	10	0.19	2.1
LP	11.0	0.17	1.3	39	0.069	63	0.072	57	0.82	82	0.10	1.7
30 m*M* NaCl												
UP	3.2	0.73	5.5	49	0.0025	0.73	0.011	2.5	0.043	4.3	0.18	4.1
LP	9.8	0.19	1.4	39	0.05	41	0.094	66	0.71	71	0.13	1.5

Note: Assay of ATPase activity was according to the method of Surowy and Sussman.[38] The assay is based on the colorimetric estimation of Pi released from ATP.[39] The assay was in a total volume of 0.5 ml for 60 min at 30°C. The reaction was started by the addition of the membrane to a mixture containing 5m*M* ATP, 5m*M* $MgCl_2$, 11 m*M* $(NH_4)_2SO_4$, 5m*M* phosphoenol pyruvate, 10μl pyruvate kinase, 10 m*M* PIPES, 100 m*M* KNO_3, 5 m*M* NaN_3, 0.1 mg/mℓ lysolecithin, adjusted to pH 6.7 with tris. Cyt.c oxidase and cyt.c reductase activities were measured by the method of Hodges and Leonard.[41] The values are from a representative experiment and are averages of duplicates. The values given for microsomes represent the amount used for partitioning with each concentration of NaCl. Typically for assays 100—200 μg of microsomes, 25—50 μg of UP and 100—200 μg of LP were used. The SA was confirmed by determining the activities at two levels of the protein. UP: upper phase, LP: lower phase. LPC: lysophosphotidylcholine.

[a] Protein determination was by the method of Markwell et al.[40]
[b] Specific activity (SA) is μmole of phosphate released/mg protein/min.
[c] Specific activity is μmole/mg protein/min.
[d] Chlorophyll was determined according to Wintermans and DeMots.[42]

Table 2
EFFECT OF NaCl ON THE PARTITIONING OF *S. COMMERSONII* PLASMA MEMBRANE IN A TWO-PHASE POLYMER SYSTEM

		ATPase activity			Cyt.c-oxidase		Cyt.c-reductase		Chlorophyll[d]		ATPase activity	
Fraction	Protein[a] (mg)	SA[b]	Fold SA	% Activity	SA[c]	% Activity	SA[c]	% Activity	(mg)	(%)	SA[b] without LPC	Activation with LPC (fold SA[b])
Microsomes	37	0.061	1	100	0.014	100	0.026	100	1.7	100	0.057	1.1
10 m*M* NaCl												
UP	8.4	0.13	2.1	48	0.0080	13	0.040	35	0.24	14	0.048	2.7
LP	16	0.050	0.83	35	0.012	37	0.023	38	0.88	52	0.017	2.9
20 m*M* NaCl												
UP	5.0	0.16	2.7	35	0.0072	7.0	0.0079	4.1	0.067	4	0.054	3.0
LP	19	0.041	0.066	33	0.0079	29	0.018	36	1.3	77	0.029	1.4
30 m*M* NaCl												
UP	3.4	0.26	4.3	38	0.0039	2.6	0	0		2.0	0.064	4.0
LP	19				0.0034	13	0.026	51	1.2	74		

Note: Details as described for Table 1.

Table 3
ATPase ACTIVITY IN PRESENCE AND ABSENCE OF DIFFERENT INHIBITORS

Inhibitor (+present −absent)	% PM-ATPase activity					
	S. tuberosum			*S. commersonii*		
	Microsomes	UP	LP	Microsomes	UP	LP
-KNO_3	115 ± 7	102 ± 1	115 ± 12	128 ± 22	100	128 ± 23
-NaN_3	130 ± 10	90 ± 0.6	126 ± 7	144 ± 24	100	115 ± 9
-Molybdate	116 ± 12	100	123 ± 7	100	100	100
+Yanadate (100 μM)	55	27	38	59	22	24

Note: Details of the assay under Table 1. The values are averages ±SD of at least three different experiments. The values presented in the table are calculated with respect to the standard assay performed with the same samples using the procedure described under Table 1 (Standard assay value taken as 100). The values for assays in presence of vanadate are from a single experiment and are averages of duplicates.

with any significant amount of mitochondrial membranes and endoplasmic reticulum. There was practically no contamination by chlorophyll. A major portion of cyt.c oxidase activity and NADH-cyt.c reductase activities and almost all the chlorophyll was recovered in the dextran-rich lower phase (LP). Hence, partitioning in presence of 30 m*M* NaCl was found to be the best for plasma membrane preparations from both the species and was employed for all further work. This procedure has been repeated a number of times and the range of different values obtained were as follows. The specific activity of ATPase in microsomes: 0.054 to 0.13 μmol Pi mg^{-1} protein min^{-1}, specific activity in the UP: 0.13 to 0.73 μmol Pi mg^{-1} protein min^{-1}, protein recovery in the UP: 5 to 10% of that in the microsomes, ATPase activity recovery in the UP: 26 to 55%. The cyt.c reductase, cyt.c oxidase, and chlorophyll content in UP was always less than 5% of that in the microsomes.

The ATPase activity in various fractions was measured in the presence and absence of lysolecithin with an objective of learning the sidedness of the membranes. Enhanced activity in presence of lysolecithin probably indicates an inward location of the ATP binding site and hence right-side-out orientation of plasma membrane vesicles. Conversely, a high ATPase activity even in the absence of lysolecithin probably indicates that ATP binding site is located on the outer surface and hence an inside-out orientation of the plasma membrane vesicles or the presence of leaky right-side-out vesicles. These results are also shown in Tables 1 and 2. The extent of activation of ATPase activity with lysolecithin obtained in UP was always higher compared to that in the LP and microsomes. This is probably due to the preferential partitioning of plasma membrane vesicles with a right-side-out orientation to the UP. In all our preparations this was a uniform feature. Lysolecithin produced a maximum of twofold activation with microsomes and LP membranes whereas in UP this ranged from four to tenfold.

In order to confirm the lack of contamination of the purified preparations by tonoplast and mitochondria, the ATPase activity was carried out in the absence of the respective ATPase inhibitors, KNO_3 and NaN_3, respectively. Contamination by nonspecific phosphatases was checked by carrying out the assay in the absence of ammonium molybdate. The specificity of the ATPase was checked by its sensitivity to vanadate. The results are shown in Table 3. In the absence of KNO_3, the total ATPase activity in the microsome and LP was 20 to 30% over that of the plasma membrane ATPase activity. This shows that there is tonoplast membrane in microsome and LP fractions. However, there was no detectable tonoplast ATPase activity in the UP membranes. Similarly, mitochondrial ATPase activity

Table 4
PLASMA MEMBRANE ATPase ACTIVITY BEFORE AND AFTER COLD ACCLIMATION OF *S. COMMERSONII* LEAVES

		Protein[a]		ATPase activity		
Treatment	**Fraction**	**(mg)**	**(%)**	**SA[b]**	**Units**	**(%)**
Nonacclimated	Microsomes	42	100	0.13	5.5	100
	UP	3.6	8.6	0.32	1.2	22
	LP					
Acclimated	Microsomes	46	100	0.21	9.7	100
	UP	3.9	8.5	0.50	2.0	21
	LP	23.8	52			

Note: Details as described under Table 1. Nonacclimated: grown at 20/15°C, day/night; Acclimated: 8-week-old acclimated plants subjected to 2 weeks 5/2°C, day/night.[43]

was present in the microsomes and LP whereas no activity was detected in the UP, further confirming the lack of mitochondrial contamination. There seems to be no phosphatase activity in any of the fractions of *S. commersonii*, but it was present in microsomes and lower phase of *S. tuberosum* and not in the UP. In the presence of 100 μM vanadate (a specific inhibitor of plasma membrane ATPase) in the UP and LP membranes there was 70 to 80% inhibition of ATPase activity and in microsomal membranes this inhibition was 55 to 60%.

2. *Effect of Freeze Thaw and Cold Acclimation on Plasma Membrane ATPase Activity*

For testing the effect and cold acclimation of freezing on plasma membrane ATPase activity, 6 to 8-week-old plants were used.[43] About 60 leaves were frozen at a rate of 1°C/hr to the desired temperature (with ice nucleation around −0.5°C). After freezing, the leaves were thawed over ice overnight and plasma membranes were prepared as described above. The total yield of the plasma membrane in terms of the total protein, total units of ATPase activity, and specific activity of ATPase were compared with that of the unfrozen and unacclimated control of the same batch of leaves.

The specific activity of plasma membrane ATPase after cold acclimation increased significantly in the leaves of *S. commersonii* (Table 4). Although the total microsomal protein was not changed significantly the specific activity of the ATPase was higher and hence the total ATPase was also higher after cold acclimation. Though the total protein in the purified plasma membrane fraction was not different in the nonacclimated and acclimated material, the specific activity and the total ATPase activity in this fraction was about 50% higher after acclimation. This indicates that there is either an increase in the amount of ATPase on the membrane or there is an enhanced kinetic efficiency of the ATPase after cold acclimation. We plan to test the ATPase activity in *S. tuberosum* (not capable of acclimation) after cold treatment. If the changes seen in *S. commersonii* after cold acclimation are not seen in *S. tuberosum* the changes could probably be related to the process of acclimation.

The results from *S. commersonii* and *S. tuberosum* are shown in Tables 5 and 6. When *S. commersonii* leaves were frozen at −3°C (non lethal stress), the total protein obtained in the microsomal fraction and the plasma membrane (UP) fraction were not significantly different compared to the protein obtained from the same membrane fraction from unfrozen control leaves (Table 5). Similarly, the specific activity of the ATPase and total ATPase activity recovered in the pure plasma membrane fractions after the −3°C freeze were not markedly different. But at −7°C (the leaves were irreversibly damaged) the amount of

Table 5
PLASMA MEMBRANE ATPase IN MEMBRANE FRACTIONS OF *S. COMMERSONII* BEFORE AND AFTER FREEZE-THAW STRESS

			Total protein[a]		ATPase activity		
Expt.	Treatment	Fraction	(mg)	(%)	SA[b]	Units	(%)
1	Unfrozen	Microsomes	29	100	0.063	1.8	100
	(control)	UP	1.4	4.8	0.62	0.87	48
		LP	13	45			
	Frozen	Microsomes	31	100	0.070	2.2	100
	(−3°C	UP	1.1	3.6	0.70	0.78	36
	nonlethal)	LP	14	45			
2	Unfrozen	Microsomes	25	100	0.093	2.3	100
	(control)	UP	2.0	8	0.38	0.76	33
		LP	9.0	36			
	Frozen	Microsomes	16	100	0.102	1.6	100
	(−7°C	UP	0.64	4	0.23	0.15	9.4
	lethal)	LP	5.9	38			
3	Unfrozen	Microsomes	43	100	0.05	2.2	100
	(control)	UP	2.3	5.4	0.38	0.89	40
	Slightly	Microsomes	33.7	100	0.077	2.6	100
	damaged	UP	1.48	4.4	0.58	0.85	33

Note: Details of the assay as in Table 1.

Table 6
PLASMA MEMBRANE ATPase ACTIVITY IN MEMBRANE FRACTIONS OF *S. TUBEROSUM* LEAVES BEFORE AND AFTER FREEZE-THAW STRESS

		Protein[a]		ATPase activity		
Temperature	Fraction	(mg)	(%)	SA[b]	Units	(%)
Unfrozen	Microsomes	15	100	0.19	2.9	100
0°C	UP	1.5	10	0.31	0.47	16
	LP	7.8	52			
Frozen	Microsomes	11	100	0.051	0.56	100
−4°C	UP	0.49	4.5	0.085	0.042	7.5
(lethal)	LP	6.10	56			

Note: Details as described under Table 1.

microsomal protein recovered from the same amount of tissue was reduced by about 35% and the total ATPase activity in the microsomal fraction was reduced by about 30%. The protein in the plasma membrane fraction represents 8% of the microsomal protein in unfrozen control, whereas it is only about 4% of the microsomal protein in the −7°C frozen sample (Table 5). Similarly, the total ATPase activity represents 33% of that in the microsomes in the unfrozen control whereas after freezing to −7°C it is only about 9% of that in the microsomes after lethal freezing.

The reduction in the total protein and also the ATPase activity in microsomes (after freezing to −7°C) indicated that there was a loss in the membrane protein which also included the ATPase. There was no noticeable change in the specific activity of ATPase in microsomes probably because both the total membrane protein and ATPase activity are reduced by similar levels. The differences between the −7°C and unfrozen leaves in the purified plasma membrane fraction with respect to total protein, specific activity, and total activity were greater than the differences in the microsomal fraction. It is possible that some of this loss could be accounted for by changes in partitioning behavior of some plasma membrane vesicles (a loss from UP to LP). However, a large portion of the loss in protein in plasma membrane preparation appears to be due to freeze-induced alterations in the plasma membrane.

The results with *S. tuberosum* frozen to −4°C (leaves were irrevesibly damaged, LT_{50} of −3°C) were similar to those obtained at −7°C with *S. commersonii* (Table 6). There was about 25% reduction in the total microsomal protein obtained after freezing but the total ATPase activity recovered in the microsomal fraction was reduced more drastically (by about 80%), and the ATPase activity recovered in the purified plasma membrane fraction was only about 10% of that in the microsomes which was only about 45% of that obtained in the unfrozen control. The specific activity of the ATPase was also reduced greatly.

It was observed during the course of above freezing studies (experiment 1 and 2, Table 5) that at a particular temperature all the leaves were not uniformly injured (injury assessed) by visual water soaking, turgidity, and ion leakage). The extent of variation differed in different batches of plants and also at different temperatures. The results obtained with such samples will be an average of all the injury levels represented in the sample. Thus, the ATPase activity measured in the membranes isolated from these samples cannot be related to a definite degree of injury. In our recent study, we separated *S. commersonii* leaves into four different categories (based on visual water soaking, turgidity and ion leakage) with defined degrees of damage representing totally reversible (fully recoverable), partially reversible, and irreversible levels of injuries. Plasma membrane was purified and ATPase activities were studied in all these categories. Cyt. c oxidase (a mitochondrial enzyme) and NADH-cyt. c reductase (an enzyme of endoplasmic reticulum activities were also examined. It was found that at reversible (slight) damage (10 to 20% water soaking, no loss of turgidity, about 20% ion leakage[10] which was about 2-fold increase in ion leakage over the control) there was 1.5 to 2-fold increase in specific activity of plasma membrane ATPase (experiment 3, Table 5). K_m was decreased 2.5- to 3-fold, at this level of injury (data not shown), reflecting an increased affinity. It is possible that this increased activity of ATPase helps in the process of recovery in slightly injured leaves. These changes might have been masked in our studies using leaves frozen to -3°C (Table 5) because of the averaging effect of some noninjured and injured leaf tissue in the sample. At irreversible damage (70 to 90% water soaking, turgidity mostly lost, about 75% ion leakage which is about 6-fold increase over the control), there was a loss of membrane protein and loss of ATPase activity (data not shown). However, even after this high degree of damage there was no effect on the activity of cyt. c oxidase and NADH-cyt. c reductase suggesting that plasma membrane ATPase is an earlier target of alteration during freezing injury. This fact is also supported by the observation by Steffen and Palta[44] that ion leakage increases before an impairment of photosynthesis and respiration becomes apparent in leaf tissue.

Results presented here provide evidence that plasma membrane ATPase is an early site of alteration following freezing injury. These alterations can be detected even in reversibly damaged tissue which is able to recover completely. Mechanism of alteration of plasma membrane ATPase following freezing injury and cold acclimation is not known. Elucidation of the critical events following freezing injury and cold acclimation necessitates studies: (1) where tissue is subjected to a realistic freeze-thaw and acclimating protocols which resemble as closely as possible those that occur in nature, (2) where tissue with as uniform a damage as possible is utilized, (3) where simultaneous investigations of in vivo and in vitro perturbation in the membrane properties can be made, and (4) where both lipid and protein components of the plasma membrane can be investigated. *Solanum* offers a good system to undertake these studies because of the availability of related species with considerable differences in freeze-stress resistance and cold acclimation ability.

V. CONCLUSIONS

Plasma membrnae was recognized as a site of freezing injury about 75 years ago. However, the nature of injury at the molecular and cellular level has been investigated only recently. Palta and co-workers have proposed that an incipient freezing injury results in an alteration of the functions of the active transport process at the cell membrane (presumably H^+-ATPase coupled transport). Subsequent to the advancement of this hypothesis, several independent investigations have provided evidence of the alterations in the activity of plasma membrane ATPase after freezing injury and cold acclimation, supporting the hypothesis. Our results with the potato species presented here also support this hypothesis.

Mechanisms by which the functions of the membrane ATPases could alter is not known. However, various possibilities exist. Some of them are (1) direct effect by alteration of conformation or functional properties, (2) an alteration of protein-lipid interaction around the ATPase site, (3) loss of coupling between ATP hydrolysis and H^+-transport, (4) loss of coupling between ATPase activity and specific transport proteins, and (5) nonavailability of ATP. At present it is not possible to distinguish between these various possibilities. It is clear that a parallel in vivo and in vitro studies are necessary to understand the definite progression of events leading to alteration of the functions of plasma membrane ATPase by freezing stress.

ACKNOWLEDGMENT

This material is, in part, based on work supported by the U.S. Department of Agriculture Competitive Grant (Agreement No. 85-CRCR-1-1673) and by the College of Agricultural and Life Sciences, University of Wisconsin, Madison.

REFERENCES

1. **Maximov, N. A.,** Chemische Schutzmittel der Pflanzen gegen erfrieren., *Ber. Deutsch. Bot. Gessel.*, 30, 52, 1912.
2. **Palta, J. P. and Li, P. H.,** Alteration in membrane transport properties by freezing injury in herbaceous plants: evidence against rupture theory, *Physiol. Plant.*, 50, 169, 1980.
3. **Palta, J. P. and Li, P. H.,** Cell membrane properties in relation to freezing injury, in *Plant Cold Hardiness and Freezing Stress: Mechanisms and Crop Implications*, Li, P. H. and Sakai, A., Eds., Academic Press, New York, 1978, 93.
4. **Palta, J. P., Jensen, K. G., and Li, P. H.,** Cell membrane alterations following slow freeze-thaw cycle: ion leakage, injury and recovery, in *Recent Advances in Plant Cold Hardiness and Freezing Stress: Mechanisms and Crop Implications*, Vol. 2, Li, P. H. and Sakai, A., Eds., Academic Press, New York, 1982, 221.
5. **Steponkus, P. L.,** Role of the plasma membrane in freezing injury and cold acclimation, *Annu. Rev. Plant Physiol.*, 35, 543, 1984.
6. **Yoshida, S.,** Studies on freezing injury of plant cells, *Plant Physiol.*, 75, 38, 1984.
7. **Hellergren, J., Widell, S., and Lundborg, T.,** ATPase in relation to freezing in purified plasma membrane vesicles from nonacclimated seedlings of *Pinus sylvestris*, *Acta Hortic.*, 168, 161, 1985.
8. **Levitt, J.,** *Responses of Plants to Environmental Stresses*, 2nd ed., Academic Press, New York, 1980.

9. **Steponkus, P. L. and Weist, S. C.,** Plasma membrane alterations following cold acclimation and freezing, in *Recent Advances in Plant Cold Hardiness and Freezing Stress: Mechanisms and Crop Implicaitons,* Li, P. H. and Sakai, A., Eds., Academic Press, New York, 1978, 75.
10. **Palta, J. P., Levitt, J., and Stadelmann, E. J.,** Freezing injury in onion bulb cells. I. Evaluation of the conductivity method and analysis of ion and sugar efflux from injured cells, *Plant Physiol.*, 60, 398, 1977.
11. **Palta, J. P., Levitt, J., and Stadelmann, E. J.,** Freezing injury in onion bulb cells. II. Post thawing injury or recovery, *Plant Physiol.*, 60, 398, 1977.
12. **Palta, J. P., Levitt, J., and Stadelmann, E. J.,** Freezing tolerance of onion bulbs and significance of freeze-induced tissue infiltration, *Cryobiology,* 14, 614, 1977.
13. **Hodges, T. K.,** ATPases associated with membranes of plant cells, in *Transport in Plants.*, Vol. 2, (Part A), Littege, U. and Pitman, M. G., Eds., Springer-Verlag, Berlin, 1976, 266.
14. **Jian, C. L., Sun, L. H., Dong, H. Z., and Sun, D. L.,** Changes in ATPase activity during freezing injury and cold hardening, in *Plant Cold Hardiness and Freezing Stress: Mechanisms and Crop Implications,* Li, P. H. and Sakai, A., Eds., Vol. 2, Academic Press, New York, 1982, 243.
15. **Pomeroy, K. M., Pihakaski, J. S., and Andrews, C. J.,** Membrane properties of isolated winter wheat cells in relation to icing stress, *Plant Physiol.*, 72, 535, 1983.
16. **Hellergren, J., Widell, S., Lundborg, I., and Kylin, A.,** Frosthardiness development in *Pinus sylvestris:* the involvement of K^+-stimulated Mg^{2+} dependent ATPase from purified plasma membrane of pine, *Physiol. Plant.*, 58, 7, 1983.
17. **Hellergren, J., Widell, S., and Lundborg, T.,** Is the plasma membrane bound ion stimulated ATPase the primary site of freezing injury?, *Plant Physiol.*, 77, 15, 1985.
18. **Ishikawa, M. and Yoshida, S.,** Seasonal changes in plasma membrane and mitochondria from Jerusalem artichoke tubers. Possible relation to cold hardiness, *Plant Cell Physiol.*, 26, 1331, 1985.
19. **Uemura, M. and Yoshida, S.,** Studies on freezing injury in plant cells, *Plant Physiol.*, 80, 187, 1986.
20. **Sze, H.,** H^+-translocating ATPases: advances using membrane vesicles. *Annu. Rev. Plant. Physiol.*, 36, 175, 1985.
21. **Heber, U., Tyankova, L., and Santarius, K. A.,** Effects of freezing on biological membranes *in vivo* and *in vitro, Biochim. Biophys. Acta.*, 291, 23, 1973.
22. **Steponkus, P. L., Evans, R. Y., and Singh, J.,** Cryomicroscopy of rye mesophyll cells, *Cryoletters,* 3, 101, 1982.
23. **Steponkus, P. L., Dowgert, M. F., and Gordon-Kamm, W. J.,** Destabilization of the plasma membrane of isolated plant protoplasts during a freeze-thaw cycle: the influence of cold acclimation, *Cryobiology,* 20, 448, 1983.
24. **Dowgert, M. F. and Steponkus, P. L.,** Behavior of the plasma membrane of isolated protoplasts during a freeze-thaw cycle, *Plant Physiol.*, 75, 1139, 1984.
25. **Carruthers, A. and Melchior, D.,** How bilayer lipids affect membrane protein activity, *Trends Biochem. Sci.*, 11, 331, 1986.
26. **Hellergren, J., Lundborg, T., and Widell, S.,** Cold acclimation in *Pinus sylvestris:* Phospholipids in purified plasma membranes from needles of pine, *Physiol. Plant.*, 62, 162, 1984.
27. **Uemura, M. and Yoshida, S.,** Involvement of plasma membrane alterations in cold acclimation of winter rye seedlings, *Plant Physiol.*, 75, 818, 1984.
28. **Yoshida, S. and Uemura, M.,** Protein and lipid compositions of isolated plasma membranes from orchard grass and changes during cold acclimation, *Plant Physiol.*, 75, 31, 1984.
29. **Arora, R. and Palta, J. P.,** Protoplasmic swelling as a symptom of freezing injury in onion bulb cells, *Plant Physiol.*, 82, 625, 1986.
30. **Cheung, W. Y., Ed.,** in *Calmodulin, Calcium and Cell Functions,* Vol. 1, Academic Press, New York, 1980, 1.
31. **Trauble, H. J.,** The movement of molecules across lipid membranes: molecular theory, *J. Membr. Biol.*, 4, 193, 1971.
32. **Levitt, J.,** The moment of frost injury, *Protoplasma,* 48, 289, 1957.
33. **Albertsson, P. A.,** Partition between polymer phases, *J. Chromatogr.*, 159, 111, 1978.
34. **Chen, H. H. and Li, P. H.,** Characteristics of cold acclimation and deacclimation in tuber bearing *Solanum* spp., *Plant Physiol.*, 65, 1146, 1980.
35. **Li, P. H. and Palta, J. P.,** Frost hardening and freezing stress in tuber bearing *Solanum* species, in *Recent Advances in Plant Cold Hardiness and Freezing Stress: Mechanisms and Crop Implications,* Li, P. H. and Sakai, A., Eds., Academic Press, New York, 1978, 49.
36. **Serrano, R.,** ATPase and proton transport in isolated plasma membranes from plants and fungi, in *Plasma Membrane ATPase of Plants and Fungi,* Serrano, R., Ed., CRC Press, Boca Raton, Fla,., 1985, 81.
37. **Kjellbom, P. and Larsson, C.,** Preparation and polypeptide composition of chlorophyll-free plasma membranes from leaves of light grown spinach and barley, *Physiol. Plant.*, 62, 501, 1984.
38. **Surowy, T. K. and Sussmann, M. R.,** Immunological cross-reactivity and inhibitor sensitivities of plasma membrane H^+-ATPase from plants and fungi, *Biochim. Biophys. Acta,* 848, 24, 1986.

39. **Dryer, L., Tammes, A. R., and Routh, J. I.,** The determinaiton of phosphorous and phosphates with N-phenyl-*p*-phenylenediamine, *J. Biol. Chem.*, 225, 177, 1957.
40. **Markwell, M. K., Haas, S. M., Beiber, L. L., and Tolbert, N. E.,** A modification of the Lowry procedure to simplify protein determinations in membrane and lipoprotein samples, *Anal. Biochem.*, 87, 206, 1978.
41. **Hodges, T. K. and Leoanard, R. T.,** Purification of plasma membrane bound ATPase from plant roots, *Methods Enzymol.*, 32, 392, 1974.
42. **Wintermans, J. F. G. M. and DeMots, A.,** Spectrophotometric characteristics of chlorophylls and their pheophytins in ethanol, *Biochim. Biophys. Acta*, 109, 448, 1965.
43. **Steffen, K. L. and Palta, J. P.** Effect of light on photosynthetic capacity during cold acclimation in a cold-sensitive and cold tolerant potato species, *Physiol. Plant.*, 66, 353, 1986.
44. **Steffen K. L. and Palta, J. P.** Photosynthesis as a key process in plant response to low temperature: alteration during low temperature acclimation and impairment during incipient freezing injury, in *Plant Cold Hardiness*, Li, P.H., Ed., Alan R. Liss, 1987, 67.

Chapter 10

CRYOPRESERVATION OF PLANT CELLS AND ORGANS

Tony H. H. Chen and P. H. Li

TABLE OF CONTENTS

I. Introduction 140

II. Importance of Cryopreservation 140
  A. Germplasm Conservation 140
  B. Preservation of Tissue Cultures 141
    1. Preservation of Morphogentic Potential 141
    2. Preservation of Biosynthetic Potential 142
    3. Preservation of Selected Mutants 142
  C. Improvement of Woody Species 142

III. Methodology of Cryopreservation 143
  A. Plant Materials 143
  B. Pregrowth Conditioning 144
  C. Cryoprotectant Treatment 145
  D. Freezing Methods 146
    1. Rapid Freezing 146
    2. Slow Freezing 147
    3. Droplet Freezing 147
  E. Storage 148
  F. Thawing 148
  G. Regrowth and Viability Testing 148

IV. Summary 149

Acknowledgment 149

References 150

I. INTRODUCTION

Cryopreservation is the preservation of viable biological samples at ultra-low temperatures (−196°C). Cellular metabolic activities are arrested by cryopreservation due to the low temperature and the unavailability of liquid water. Consequently, little or no biochemical changes occur with time. In principle, it is similar to the practice of storing perishable food in the frozen state. In cryopreservation, however, the biological sample must not be killed by the low temperature, thus allowing subsequent regeneration.

The development of the cryopreservation technology began in 1949 when it was discovered glycerol protected living cells from freezing injury.[1] In 1959, dimethylsulfoxide (DMSO) was discovered to be one of the most effective cryoprotectants known for both animal and plant tissues.[2] Earlier cryopreservation studies were carried out almost entirely on animal cells or tissues. These methods laid the ground work for plant cryoprotection. In plant systems, successful cryopreservation was not reported for cultured plant cells until 1968,[3] and until 1976 for meristem tissues followed by regeneration of whole plants.[5] Cryogenic storage of seeds, which generally does not require the use of cryoprotectants, dates back to the late 1800s.[5]

In the past two decades, considerable effort has gone into the establishment of cryopreservation protocols for cells and organs of a wide range of plant species.[5] These protocols rely on the identification of suitable cryoprotectants and on suitable cooling and warming conditions for maximal recovery. Ironically, "the science of cryobiology has not advanced significantly during the past two decades".[6]

Cryopreservation of valuable germplasm of cultivated plant species and their wild relatives has received wide attention. The use of in vitro methods, for long-term storage of plant germplasm and for international exchange of clonal propagated plant species, has great advantages over conventional methods. The readers are referred to a recently published book[5] and other review articles[7-12] for practical details of cryopreservation of plant materials.

II. IMPORTANCE OF CRYOPRESERVATION

A. Germplasm Conservation

The collection and conservation of germplasm are necessary to maintain a broad gene base for crop improvement. The irretrievable loss of naturally occuring germplasm is a major concern of plant breeders throughout the world. The loss of plant germplasm is attributed to (1) the destruction of natural habits for alternative uses; (2) the introduction of high yielding varieties and the abandonment of traditional varieties rich in genetic variability[13] and (3) human negligence. The preservation of valuable germplasm for future use in plant improvement is imperative.

Seed storage has been the most common practice for maintaining plant germplasm throughout our agricultural history. For plant species in which the germplasm can be maintained as seeds, cold storage at −20°C with a seed moisture content of 5 to 7% greatly extends the seed viability. To prevent physiological deterioration and loss of viability that can occur during long-term storage, preservation in liquid nitrogen is the preferred method.[5]

The procedures used for seed storage cannot be applied directly to the preservation of all plant species. Some species produce seeds that cannot be dried without immediate injury and viability loss, (i.e., recalcitrant seeds).[5] In some plants which are genetically heterozygous, propagation through seeds results in the loss of unique genotypes. Plants which do not produce seeds must be stored in a vegetative form such as tubers, roots, cuttings, etc. Maintaining vegetatively propagated plants in the field is labor intensive, expensive, and often results in the loss of valuable genetic material due to disease, pests, climatic perturbations, natural disasters, and human error.[10-11] This resulted in a need to develop cryopreservation techniques suitable for in vitro plant genetic conservation.

Table 1
CRYOPRESERVATION OF PLANT MERISTEMS AND SHOOT TIPS

Plant Species	Material[a]	Ref.
Arachis hypogaea	M	51
Brassica napus	ST	45
Cicer arietinum	M	14, 51
Dianthus caryophyllus	ST	4, 56
Fragaria × ananassa	M	30
Fragaria × ananassa	ST	27
Malus domestica	ST	25
Manihot esculenta	M	31
Pisum sativum	M	29
Solanum etuberosum	ST	57
S. goniocalyx	ST	58
S. tuberosum	M	50
S. tuberosum	ST	58

[a] M = meristems; ST = shoot tips.

Meristem culture, a technique that facilitates the recovery of pathogen-free, genetically stable progeny, is widely used for mass propagation of several crop species. These advantages prompted the use of meristem/shoot tips for in vitro preservation of germplasm. Cryopreservation of meristems or shoot tips is one of the best approaches for the long-term preservation of vegetatively propagated crops and plants which produce recalcitrant seeds.[5,14] Meristem/shoot tips can be stored for long periods in a sterile, stable environment without chromosomal aberrations occurring. Examples of cryopreservation of meristem/shoot tips for some crop species are summarized in Table 1.

B. Preservation of Tissue Cultures

Maintenance of cell cultures in vitro for long periods may generate several problems. Currently, repeated subculture is the most common method of maintaining cell cultures. Subculturing over an extended period of time may induce undesirable changes such as chromosomal aberrations or the loss of biochemcical characteristics and morphogenetic potentials. Cryopreservation overcomes many of these problems for a wide spectrum of plant tissue cultures, ranging from protoplasts to regenerant plantlets. Plant cell cultures which have been successfully stored in liquid nitrogen (LN) and which retained viability after thawing are discussed in several reviews.[7,8, 10-12] Chen and Kartha[7] categorized three important uses of cryopreservation of plant cell cultures (Table 2).

1. Preservation of Morphogenetic Potential

The feasibility of applying plant cell tissue culture technology to crop improvement relies on in vitro plant regeneration. Many plant tissue cultures decline or lose regeneration potential and increase in developmental and genetic anomalies during repeated subculture. From the limited information available, it appears that plants regenerated from cryopreserved cells are able to maintain the karyotypes of the original cultures. Nag and Street[15] examined large numbers of metaphase chromosomes of carrot plants regenerated from cryopreserved cells and found no evidence of change in karyotype. Baja[16] also found that plants regenerated from cryopreserved haploid tobacco plants showed a normal haploid chromosome number (n = 24).

Table 2
CRYOPRESERVATION OF IN VITRO MORPHOGENETIC AND BIOSYNTHETIC POTENTIALS, AND SELECTED MUTANTS OF PLANT CELLS

Objective	Plant species	Cultures	Ref.
Morphogenetic potential	*Daucus carota*	Suspension	15
	Hyoscyamus muticus	Suspension	47
	Lavandula vera	Callus	59
	Nicotiana tabacum	Haploid suspension	16
	Medicago sativa	Callus	60
	Phoenix Dactylifera	Callus	61
	Triticum aestivum	Callus	37
	Ulmus americana	Callus	62
Biosynthetic capability	*Catharanthas roseus*	Alkaloid producing cell lines	19
	Daucus carota	Suspension	20
	Digitalis lanata	Suspension	21
	Dioscorea deltoidea	Suspension	63
	Lavandula vera	Green callus	57
Selected mutants	*Daucus carota* and *Nicotiana tabacum*	Amino acid resistant cell lines	64

2. Preservation of Biosynthetic Potential

Plant cell cultures are potential sources for the production of specific chemicals that are also produced in the plants from which they were derived. The combination of plant cell culture techniques with fermentation technology for commercial production of specific compounds has been well documented.[17] Cultures which produce such compounds in high yields are often biochemically unstable to repeated subculturing. For example, the production of alkaloids from *Hyocyamus niger* cell lines gradually changes in both total alkaloid content and the species of alkaloids after years of subculture.[18] Cryopreservation is used to maintain cell strains which produce compounds of medical or industrial importance in high yield. A procedure for the preservation of *Catharanthus roseus* cell lines which produces high levels of important alkaloids has been developed.[19] Similarly, 25 different cultures of wild carrot produced approximately the same quantity of anthocyanins after cryogenic storage as did the unfrozen controls.[20] Cryopreserved cells of *Digitalis lanata* also retained the biochemical activity relative to the transformation of cardenolides.[21]

3. Preservation of Selected Mutants

The selection of mutant cell line in vitro has provided invaluable materials for studying the biochemical, physiological, and genetic nature of plants. In vitro selection techniques have also resulted in the development of disease[22] and herbicide-resistant[23] plants. Recently, increasing interest in the genetic engineering of plant species makes it important to preserve cultured cells with unique characteristics. However, the maintenance of the selected mutant or genetic engineered cell lines is problematic as conventional repeated subculture often results in the reversion to wild types. Cryopreservation provides a means for preserving those cell lines.

C. Improvement of Woody Species

The application of tissue culture/genetic engineering techniques to the improvement of many woody species is hampered by poor plant regeneration from mature tissue, or cultures derived from explants of mature plants. The application of cryopreservation techniques to tree improvement or enhancement of forest production has been proposed.[8] By generating plants from mature tissues, it may be possible to select superior trees from forests or orchards. These superior clones could be propagated by in vitro methods for large scale plantation for use in reforestation. However, the technology for routine regeneration of plants from mature tissue is not well developed.

The cryogenic storage of juvenile tissues would bypass the difficulties of regeneration of plants from mature tissues. Long-term storage of juvenile clones allows the researchers to perform field evaluation of the same clones. At the end of the test period (while the clones in the field have passed their juvenile stage and lost their potential for in vitro cloning), the investigators could use cryopreserved juvenile clones and multiply them for nurseries.

Plants originated from hybrid seeds by controlled pollination generally perform better than the progeny of open-pollinated seeds. Such an operation is very costly and it often takes many years to collect enough seeds for a plantation.[24] An alternative would be to produce clones from limited supplies of hybrid seeds. Cryogenic storage of this material either as zygotic embryos, seeds, or in vitro induced buds or meristems, would provide a continous supply of superior clones of desirable genotypes. Cryopreservation could be employed as a versatile tool for preserving existing natural variability in germplasms and also for conserving the variability induced in vitro through biotechnical manipulations.

III. METHODOLOGY OF CRYOPRESERVATION

Cryopreservation of biological material involves reduction of metabolic activities by lowering the sample temperature to that of liquid nitrogen (−196°C). The reduction in metabolic activities minimizes possible genetic changes during the period of storage. Unfortunately, not all plant materials can survive cyrogenic storage.

Very few plant materials can be stored directly in LN without the use of cryoprotectants. Very winter hardy woody plant materials, when sufficiently dehydrated by extracellular freezing, can survive direct immersion to LN temperature if first precooled to −30 or −40°C. Winter vegetative buds of several fruit trees prefrozen to temperatures ranging from −30 to −50°C, have survived immersion in LN.[25-27] Each cryopreservation protocol is actually a series of specific operations prior to, during, and after freezing and thawing. The judicious use of cryoprotectants is critical in the development of a cryopreservation protocol.

Freezing of an aqueous solution is a complex process even in a relatively simple solution. The complexity of the freezing process is further compounded in a multicomponent system, such as plant organs and living cells. For detailed information concerning the freezing injury and resistance of plant cells, readers are referred to several excellent books.[5,6,28]

A. Plant Materials

In a broad sense, any part of a plant of culture can be the target material for cryopreservation. For germplasm preservation, however, it must be possible to regenerate plants from cryopreserved material. This limits the options for choosing plant cells or organs for cryopreservation. Table 3 lists the choice plant organs or cultures for cryopreservation. For most self-pollinated plant species, storing seeds in LN is a practical means for long-term storage of plant germplasm. The details for cryopreservation of pollen, embryos, and seeds can be found elsewhere[5] and will not be discussed here.

As mentioned previously, a high frequency of plant regeneration from in vitro cultured meristems of shoot tips can be achieved in many plant species. Various protocols regarding cryopreservation of meristems have been outlined by Kartha.[5,9,14,29-31] Very winter hardy woody plants and dormant prefrozen winter vegetative buds of several fruit trees (apple, gooseberry, currant, raspberry, and pear) can be cryopreserved directly in LN.[25,26] Even

Table 3
REGENERATION AND RISK OF GENETIC CHANGE OF SEVERAL PLANT MATERIALS CRYOPRESERVED

Plant material	Means of regeneration	Risk of genetic change
Seeds	Germination	Low
Meristems/shoot-tips	In vitro culture	Low
Pollens	Pollen culture	High
	Pollination	High
Anthers	Anther culture	High
Embryos (zygotic or	Direct germination invitro	Low
somatic embryos	Callusing followed by plant regeneration	High
Twigs	Cutting	Low
	Grafting	Low
Buds	Grafting	Low
Suspension and callus culture	In vitro regeneration	High

herbaceous plant meristems, such as carnation shoot tips,[32] can survive LN immersion providing certain appropriate cryoprotective treatments are made.

In suspension and callus cultures, cells exhibit a variety of physiological and morphological variation. The ability of these cells to withstand the freezing and thawing stress varies, and only a portion of the total cell population survives the cryopreservation treatment. In many cases, high cell viability following cryopreservation was obtained for cells at late lag or early division phases.[33-35] Cells in early lag or stationary phases are very sensitive to freezing, due to large cell size, high water content, and large vacuoles of cells in this phase of development. The dense, nonvacuolated cytoplasm in predominantly small sized cells at late lag or early division phase may contribute to enhanced survival.[36-38]

B. Pregrowth Conditioning

Preconditioning of plant material prior to cryopreservation has been shown to increase the viability of certain species. Cold treatment (4°C for 3 days) of clonal carnation plants prior to cryopreservation dramatically increased the survival of shoot tips.[32] Supplementation of the culture medium with osmotically active compounds such as mannitol, sorbitol, sucrose, and proline resulted in an increase in freezing resistance. Preculture of *Acer pseudoplatanus* and *Capsicum annuum* cells in a mannitol enriched medium reduced the cell volume and increased postfreezing survival.[12] However, this treatment did not enhance the survival of *Nicotiana sylvestris* cells,[39] alkaloid producing cell cultures of *C. roseus,*[19] and soybean.[40] Pritchard et al.[40-42] reported that certain physiological and biochemical changes occur in both sycamore and soybean cells when cultured in manitol enriched medium, but the two species displayed differential survival following post-thaw cycle. These results suggest the existence of species-specific differences in their physiological responses to identical pregrowth regimes.

Preculturing cells or meristem in a DMSO-supplemented medium for a period of 24 to 48 hr prior to freezing has resulted in enhanced survival.[29-31,37] However, DMSO at concentrations exceeding 5% may be cytotoxic to certain species or cell lines. For example, preculturing the nonalkaloid producing *C. roseus* cells for 24 hr in medium supplemented with 5% DMSO resulted in 34% viability loss prior to cryopreservation. However, cells which survived the DMSO treatment had a much higher survival rate following cryopreservation than non-DMSO treated cells.[38] A similar method was not applicable to the cryopreservation of three cell lines of the alkaloid producing *C. roseus.*[19]

Table 4
RECLASSIFICATION OF PLANT CRYOPROTECTIVE COMPOUNDS

Cryoprotectants	Old Classification	New classification	Permeability[a] Cell wall	Permeability[a] Plasmalemma
DMSO	Permeable	Category 1	+	+
Glycerol	Permeable	Category 1	+	+
Sucrose	Nonpermeable	Category 2	+	−
Glucose	Nonpermeable	Category 2	+	−
Manitol	Nonpermeable	Category 2	+	−
Sorbitol	Nonpermeable	Category 2	+	−
Proline	Nonpermeable	Category 2	+	−
PEG 1,000	Nonpermeable	Category 2	+	−
PEG 6,000	Nonpermeable	Category 3	−	−
PVP 40,000	Nonpermeable	Category 3	−	−

[a] + = permeable; − = nonpermeable.

C. Cryoprotectant Treatment

The application of cryoprotectants, singly or in combination, is essential for cryopreservation of many noncold hardy plant materials.[5] Cryoprotectants protect living cells from the deleterious effects of freezing and thawing by a mechanism not yet understool. Traditionally, cryoprotectants are categorized as permeating or nonpermeating with respect to the plasma membrane. It should be noted that this classification was based on animal cells which do not have a cell wall. The most commonly used permeating additives are DMSO and glycerol. Nonpermeating compounds include sugars, sugar alcohols, and high molecular weight additives, such as polyvinylpyrolidone (PVP), polyethylene glycol (PEG), dextran, and hydroxyl starch. The mechanism by which cryoprotective compounds protect living cells against freezing injury has not been well understood.[5]

Recently, Tao and Li[43] reclassified plant cryoprotectants into three categories:

1. Those that penetrate both the cell wall and the plasmalemma, e.g., glycerol and DMSO, etc.
2. Those that penetrate the cell wall, but not the plasmalemma, e.g., sucrose, mannitol, prolines, and polymers with mol wt less than 1000 dalton, etc.
3. Those that penetrate neither the plasmalemma nor the cell wall, e.g., soluble proteins, polysaccharides, PEG 6000, PVP, etc.

Table 4 lists some of the more frequently used cryoprotectants for plants.

Where application of a single cryoprotectant does not result in high survival, a mixture of cryoprotectants has been shown to result in enhanced survival. Ulrich et al.[44] demonstrated the successful cryopreservation of sugar cane callus using a mixture of cryoprotectants involving PEG, glucose, and DMSO (P-G-D, 10-8-10%). In subsequent studies, several other cell or callus lines have been cryopreserved using the same mixture.[5] Prior to the discovery of a sorbitol-DMSO system,[19] the most effective cryoprotectant mixtures applicable to a variety of cell cultures were 0.5 *M* DMSO + 0.5 *M* glycerol + 1 *M* sucrose, and 0.5 *M* DMSO + 0.5 *M* glycerol + 1 *M* proline.[5]

According to Tao and Li,[43] cryoprotectants act at different sites and thus have different cryoprotective roles. Cryoprotectants which act at different sites or have different roles when combined often enhanced survival compared to a combination of cryoprotectants from the same category. For example, the cryoprotectant mixture used by Ulrich et al.[44] include

DMSO or glycerol (Category 1), glucose (Category 2), and PEG 6000 (Category 3). The mixture used by Withers and associates[5,10-12,35,45-47] include DMSO or glycerol (Category 1), sucrose or proline (Category 2). The mixture used by Chen et al.[19] for the cryopreservation of *C. roseus* alkaloid producing cell lines includes DMSO (Category 1) and sorbitol (Category 2). Cryoprotectants of Category 1 protect protoplasts from the freeze-induced excess dehydration by reducing the amount of freezable water.[36,43] Cryoprotectants of Category 2 form a buffer layer between the cell wall and the plasmalemma, thus protecting the outer surface of plasmalemma from mechanical stress.[43] Compounds from Category 3, e.g., PEG 6000, form a second buffer layer next to the outer surface of the cell wall, thus mitigating the mechanical stress of the growing ice in protoplasts. The interpretation fits well with the currect hypothesis of freezing injury/resistance, and is supported by nuclear magnetic resonance (NMR) studies.[36]

Sorbitol has been used either as an osmoticum in the preculture medium or as a cryoprotectant during the freezing stress. Weber et al.[48] successfully cryopreserved cell cultures of *Glycine max, Datura innoxia, Brassica napus* and *Daucus carota* and cell culture-derived protoplasts of *G. max* using this technique. The cell cultures of the alkaloid-producing *C. roseus* could not be cryopreserved using sorbitol as the only cryoprotectant.[19] If the cells were precultured in a nutrient medium containing 1 *M* sorbitol for 6 to 20 hr, however, they could then be cryopreserved using 1 *M* sorbitol + 5% DMSO as a cryoprotectant.

To determine the mode of action of cryoprotectants, the freezing characteristics of solutions of DMSO and sorbitol and the freezing characteristics of the alkaloid producing *C. roseus* cells treated with DMSO and/or sorbitol, were examined by NMR and differential thermal analysis (DTA).[36] DMSO or sorbitol in the liquid medium affected the temperature of nucleation and the rate of ice crystallization. Less water crystallized at temperatures below −30°C in DMSO-treated cells as compared to controls. Similar results were obtained with sorbitol-treated cells, except sorbitol had less effect on the amount of water crystallized below −25°C. There was a close association between the percent unfrozen water at −40°C and percent cell survival after freezing in liquid nitrogen. These studies indicate that in the alkaloid producing *C. roseus* cells, the amount of liquid water at −40°C is critical in cryopreservation and the combination of DMSO and sorbitol was effective in preventing water from crystallizing.[36]

D. Freezing Methods

Freezing methods employed in cryopreservation of plant cells and organs fall into three major categories.

1. Rapid Freezing

Rapid freezing is generally accomplished by direct immersion of the specimens in liquid nitrogen. The cooling rate is in the order of several hundred to several thousand degrees (°C) per min, depending on the size of the sample and the container used to host the sample. For example, cooling rates of approximately 2500°C/min have been achieved by dispensing a small amount of sample in ministraw.[49] Under such a rapid cooling condition, it is likely that intracellular freezing would occur as the rate of water outflow is not fast enough to reach equilibrium with the extracellular ice crystals. Thus, many small ice crystals may be formed inside the cytoplasm instead of forming large ice crystals as in the case of lethal intracellular freezing. Alternatively, the ultra-rapid cooling may result in vitrification, (i.e., glass transition of water) of cellular water[6,49] and result in a high rate of survival after cryopreservation.[49]

Rapid freezing methods have been used to cryopreserve shoot tips of carnation,[4,32] potato,[50] rape,[65] peanut and chickpea,[51] and strawberry.[27] Most of the studies in cryopreservation using rapid freezing methods have been done with cryoprotectant-treated meristems or shoot tips, or extremely hardy plant cells.[25,27] This method is generally unsuitable for the cryopreservation of cell cultures.

2. Slow Freezing

When plant cells are exposed to a progressive temperature reduction, supercooling may occur, but eventually extracellular ice forms. It is belived that the plasma membrane acts as a barrier that prevents the growth of external ice into the supercooled cytoplasm.[52] Cells lose water due to the vapor pressure gradient created by the extracellular ice.[53] If the cooling rate is slow, water outflow is in equilibrium with the development of the vapor pressure gradient resulting in a progressive loss of cytoplasmic water. At a slow cooling rate, freeze-induced dehydration steadily lowers the freezing point of the cell sap, thus preventing the cell from intracellular ice formation. Upon extracellular freezing, the cytoplasm will be effectively concentrated and plant cells survive if adequately dehydrated.

Three ways of achieving protective dehydration are (1) preincubation in hypertonic solution, (2) treatment with penetrating cryoprotective chemicals, and (3) slow cooling at constant or varing cooling rates, or holding the samples at one or more intermediate sub zero temperatures. The degree of dehydration is dependent upon the time a sample is held at a sub zero temperature. Some samples achieve freezing equilibrium within minutes whereas for other samples hours are required for freezing to reach equilibrium. The most appropriate holding temperature is determined by the physiological state of the specimen, the pregrowth conditions, and the cryoprotectant used.[10,11]

The success of the slow-freezing method depends on cooling rate, pretreatments and cryoprotection, physiological state of the experimental material, and the terminal freezing temperature (the temperature prior to immersion in liquid nitrogen).

In most cryopreservation protocols, plant cells are immersed in a cryoprotectant solution. The following conditions have been shown to influence the survival of cells cryopreserved:

1. Cooling rate prior to the ice nucleation of external solution
2. The ice nucleation temperature of the external solution
3. Cooling rate of the frozen external solution
4. The eutectic temperature of the solution
5. The length of time the sample is held isothermal at the prefreezing temperature prior to immersion in liquid nitrogen

For practical convenience, most reseachers use linear constant cooling rate throughout the cooling process prior to immersion in liquid nitrogen. It is not known how survival would be affected if the cooling rate was nonlinear or decreased stepwise.

Prior to immersion in liquid nitrogen, a cooling rate of 0.5 to 2°/min to terminal temperatures between −30 to −40°C is commonly used. The optimal cooling rate depends on the type of cells, the cryoprotectant used, and the preculturing methods. Each plant specimen is characterized by its unique narrow range of optimal cooling rate. Cells may be immersed in liquid nitrogen upon reaching the terminal prefreeze temperature or held at this prefreeze temperature for a fixed period. It is thought that cells can be stored in liquid nitrogen indefinitely.

3. Droplet Freezing

In the droplet freezing method, developed by Kartha et al.[31] for preservation of cassava meristems, the cryoprotectant solution is dispensed in droplets of 2 to 3 $\mu\ell$ on an aluminum foil sheet contained in a petri dish. Each meristem is transferred to a droplet and frozen 0.5°C/min to a terminal temperature ranging from −20 and −40°C (depending on the species or culture line) prior to immersion in liquid nitrogen. This method has a high survival rate because aluminum foil is an efficient thermal conductor allowing rapid cooling of the droplet, and the heat of fusion of the small droplets would be very small compared to larger volumes used in conventional freezing. Only a small amount of extracellular ice forms from the small droplets, thus reducing the mechanical effects of the growing ice crystal. The droplet freezing technique has been successfully applied to the cryopreservation of both cassava meristems[31] and cultured cells[65] which are difficult to cryopreserve when frozen in glass or plastic cryogenic ampules.

E. Storage

From a physicochemical point of view, a compounded system must be prevented from reaching its state of minimum free energy in order to maintain a delicate balance within the system.[6] In an aqueous system at subfreezing temperature, where ice and liquid water coexist, ice is the stable phase. Thus, ice crystals tend to grow at the expense of liquid water. To achieve long-term storage at subfreezing temperatures, one must prolong the thermodynamically metastable state. Temperature fluctuation, even for a short period, during cryogenic storage may result in failure.

Cells cooled below their eutectic point are subjected to recrystallization or grain growth of ice crystals during storage. Small ice crystals have higher surface free energies than large ice crystals. Small ice crystals reduce their surface energies by either growing or by fusing with other small crystals.[52] Recrystallization occurs in pure water at temperatures as low as −100°C, and occurs in gels and concentrated aqueous solutions at higher temperatures.[52] The rate depends on the size of the crystals formed during cooling and the subsequent temperatures at which they are maintained. The rate is higher in smaller crystals and at higher temperatures. The normal storage temperature of −196°C prevents the change or growth of ice crystals during storage. Little information is available on the effects of higher storage temperature on retention of viability.

Most reports on cryopreservation of plant material in liquid nitrogen deal only with storage for a short period of time. The effects of long-term cryogenic storage on the post-thawing plant regeneration have been reported for only a few species up to a few weeks or a few months.[5,51] It is generally believed that if the specimen can survive liquid nitrogen for the first few minutes, it will be able to survive prolonged storage.

F. Thawing

The events in thawing do not simply reflect freezing in reverse. During thawing, changes in ice crystal structure have been shown to occur which result in reduced viability. Caution must be taken to avoid recrystallization and ice crystal growth during thawing. In most cryopreservation studies, rapid thawing is achieved by immersing the specimen containers for 1 to 2 min in a 35 to 40°C water bath or directly immersing the specimens in a washing solution heated to 35 to 40°C until completely thawed.

In some cases, slow warming is as effective as rapid thawing but never superior.[34] This is in contrast to cold hardiness studies with intact plants or plant parts, which show that slow warming (thawing) generally gives maximal survival.[28] One must keep in mind that the cooling rates used by the cryopreservation practitioners are from approximately 0.5 to 4000°C/min, whereas cooling rates used by cold hardiness researchers rarely exceed 10°C/hr.

G. Regrowth and Viability Testing

In cryopreservation, the objectives have been re-establishment of the culture or the regeneration of whole plants. The factors which affect post-thaw regrowth include the removal of cryoprotectants, post-thaw culture conditions, and culture regrowth medium.

Upon thawing, several cryoprotective additives applied prior to freezing are potentially toxic to the plant cells. Thus, immediate post-thaw removal of the cryoprotectants has been prescribed in most protocols.[15,19,38,54,55] In contrast to the above, Withers and King[46] found that suspension cultures of *Zea may* recovered only when layered over a semisolid medium without washing. Washing at 0°C, a common procedure, was found to reduce the viability of cryopreserved rice and sugar cane cells[55] and other cell cultures.[19,48] Chen et al.[19] plated nonwashed thawed cells of *C. roseus* onto filter paper over an agar medium for 4 to 5 hr to allow the cryoprotectants to diffuse into agar medium. This simplified the post-thaw handling, and also provided means for removing the toxic effects of cryoprotectants. In those cell lines, about 60% of cells were viable immediately after thawing and prior to washing. If washed according to the procedures for nonalkaloid producing cell line No. 916,[38] the viability was less than 10% and no regrowth was observed after plating on filter paper discs. The regrowth was observed when nonwashed cells were plated onto filter paper over agar medium for 4 to 5 hr and then transferred to fresh medium.

Depending on the material chosen for cryopreservation, various tests are used to assess viability. The most reliable test for assessing viability of cryopreserved cells is regrowth which unfortunately is the most time consuming. Viability is determined by the regrowth of the meristems/shoot-tips into shoots or plantlets. However, very often other forms of regrowth such as callus or leaves are observed.[5,31]

Rapid viability tests using vital stains, such as the triphenyltetrazolium chloride, fluorescein diacetate, and Evan's blue estimate viability within hours.[10,11] Each rapid viability test method has certain limitations; therefore, a combination of vital assays is required. In some cases, high TTC or FDA readings have been observed immediately after thawing while no regrowth could be obtained from the same samples. Conversely, samples with very low TTC readings have produced vigorous regrowth.[8] These differences between viability tests and regrowth may be related to the varying ability of the cultures to repair damage encountered during freezing and thawing.

IV. SUMMARY

Cryopreservation techniques provide a means of preserving valuable germplasm for a long period of time under stable conditions. Cryopreservation of plant cell cultures is used for the preservation of morphogenetic potential, biosynthetic capability, and selected mutants. Protocols have been developed for the cryogenic storage of many plant cells, tissues, and organs of many plant species. However, almost all of the protocols have been developed from an empirical approach. It is still not known how cryoprotective compounds protect plant cells against freezing injury, where their site(s) of action are located, or why some species or cultures can be readily cryopreserved, while others are recalcitrant to cryopreservation. The ultimate goal is the preservation of cells, tissues, or organs of all species for an indefinite period of time. Undoubtedly, further improvement in cryopreservation methodology to achieve such a goal depends on the answers to the above questions.

ACKNOWLEDGMENT

Oregon Agricultural Experiment Station Technical Paper No. 8123.

REFERENCES

1. **Polge, C., Smith, A. V., and Parkes, A. S.,** Revival of spermatozoa after vitrification and dehydration at low temperature, *Nature (London)*, 164, 666, 1949.
2. **Lovelock, J. E. and Bishop, M. W. H.,** Prevention of freezing damage to living cells by dimethylsulfoxide, *Nature (London)*, 183, 1394, 1959.
3. **Quatrono, R. S.,** Freezing-preservation of cultured flax cells utilizing DMSO, *Plant Physiol.*, 43, 2057, 1968.
4. **Siebert, M.,** Shoot initiation from carnation shoot apices frozen to −196°C, *Science*, 191, 1178, 1976.
5. **Kartha, K. K.,** *Cryopreservation of Plant Cells and Organs*, CRC Press, Boca Raton, Fla., 1985.
6. **Franks, F.,** *Biophysics and Biochemistry at Low Temperatures*, Cambridge University Press, Cambridge, 1985.
7. **Chen, T. H. H. and Kartha, K. K.,** Cryopreservation of plant cells and organs, in *Forest and Crop Biotechnology — Progress and Prospects*, Valentine, F. A., Ed., Springer-Verlag, New York, 1987, chap. 14. (In Press).
8. **Chen, T. H. H. and Kartha, K. K.,** Cryopreservation of woody species, in *Cell and Tissue Culture in Forestry*, Bonga, J. and Durzan, D. J., Ed., Martinus Nijhjoff, New York, 1987, 305.
9. **Kartha, K. K.,** Cryopreservation of germplasm using meristem and tissue culture, in *Application of Plant Cell and Tissue Culture to Agriculture and Industry*, Tomes, D. T., Ellis, B. E., Harney, P. M., Kasha, K. J., and Peterson, R. L., Eds., University of Guelph, Ontario, 1982.
10. **Withers, L. A.,** *Tissue Culture Storage for Genetic Conservation*, Tech. Rep. Int. Board Plant Genetic Resources, FAO, United Nations, Rome, 1980.
11. **Withers, L. A.,** Low temperature storage of plant tissue cultures, in *Advances in Biochemical Engineering*, Fiechter, A., Ed., 11, 102, Springer-Verlag, Berlin, 1980.
12. **Withers, L. A. and Street, H. E.,** Freeze-preservation of plant cell cultures, in *Plant Tissue Culture and Its Bio-technological Application*, Barz, W., Reinhard, E., and Zenk, M. H., Eds., Springer-Verlag, Berlin, 1977, 226.
13. **Plucknett, D. L., Smith, N. J. H., Williams, J. T., and Anishetty, N. M.,** Crop germplasm conservation and developing countries, *Science*, 220, 163, 1983.
14. **Kartha, K. K. and Gamborg, O. L.,** Meristem culture techniques in the production of disease-free plants and freeze-preservation of germplasm of tropical tuber crops and grain legumes, in *Diseases of Tropical Food Crops*, Maraite, H. H. and Meyer, J. A., Eds., Universite Catholique, Louvain-la-neuve, Belgium, 1978, 267.
15. **Nag, K. K. and Street, H. E.,** Carrot embryogenesis from frozen cultured cells, *Nature (London)*, 245, 270, 1973.
16. **Bajaj, Y. P. S.,** Regeneration of plants from cell suspensions frozen at −20, −70 and −196°C, *Physiol. Plant.*, 37, 263, 1976.
17. **Zenk, M. H.,** The impact of plant cell culture on industry, in *Frontiers of Plant Tissue Culture*, Thorpe, T. A., Ed., Int. Assoc. Plant Tissue Culture, University of Calgary, 1978, 1.
18. **Dhoot, G. K. and Henshaw, G. G.,** Organization and alkaloid production in tissue cultures of *Hyoscyamus niger*, *Ann. Bot.*, 41, 943, 1977.
19. **Chen, T. H. H., Kartha, K. K., Leung, N. L., Kurz, G. W. G., Chatson, K. B., and Constabel, F.,** Cryopreservation of alkaloid-producing cell cultures of periwinkle *(Catharanthus roseus)*, *Plant Physiol.*, 75, 726, 1984.
20. **Dougall, D. K. and Whitten, G. H.,** The ability of wild carrot cell cultures to retain their capacity for anthocyanin synthesis after storage at −140°C, *Planta Med. Suppl.*, 129, 1980.
21. **Seitz, U., Alfermann, A. W., and Reinhard, E.,** Stability of biotransformation capacity of *Digitalis lanata* cell cultures after cryogenic storage, *Plant Cell Repts.*, 2, 273, 1982.
22. **Gengenbach, B. G. and Green, C. E.,** Selection of T-cytoplasm maize callus cultures resistant to *Helminthosporium maydis* race T pathotoxin, *Crop Sci.*, 15, 645, 1975.
23. **Chaleff, R. S. and Parsons, M. F.,** Direct selection in vitro for herbicide-resistant mutants of *Nicotiana tabacum*, *Proc. Natl. Acad. Sci. U.S.A.*, 75, 5104, 1978.
24. **Powledge, T. M.,** Biotechnology touches the forest, *Biotechnology*, 2, 763, 1984.
25. **Katano, M., Ishihara, A., and Sakai, A.,** A survival of dormant apple shoot tips after immersion in liquid nitrogen, *HortScience*, 18, 707, 1983.
26. **Sakai, A. and Nishiyama, Y.,** Cryopreservation of winter-vegetative buds of hardy fruit trees in liquid nitrogen, *HortScience*, 13, 225, 1978.
27. **Sakai, A., Yamakawa, M., Sakata, K., Harada, T., and Yakuwa, T.,** Development of a whole plant from an excised strawberry runner apex frozen to −196°C, *Low Temp. Sci. Ser. Bull.*, 36, 31, 1978.
28. **Levitt, J.,** *Responses of Plants to Environmental Stresses*, Vol. 1, 2nd ed., Academic Press, New York, 1980.

29. **Kartha, K. K., Leung, N. L., and Gamborg, O. L.**, Freeze-preservation of pea meristems in liquid nitrogen and subsequent plant regeneration, *Plant Sci. Lett.*, 15, 7, 1979.
30. **Kartha, K. K., Leung, N. L., and Pahl, K.**, Cryopreservation of strawberry meristems and mass propagation of plantlets, *J. Am. Soc. Hort. Sci.*, 105, 481, 1980.
31. **Kartha, K. K., Leung, N. L., and Mroginski, L. A.**, *In vitro* growth responses and plant regeneration from cryopreserved meristems of cassava *(Manihot esculenta* Crantz), *Z. Pflanzenphysiol.*, 107, 133, 1982.
32. **Siebert, M. and Wetherbee, P. J.**, Increased survival and differentiation of frozen herbaceous plant organ cultures through cold treatment, *Plant Physiol.*, 59, 1043, 1977.
33. **Sala, F., Cella, R., and Rollo, F.**, Freeze-preservation of rice cells, *Physiol. Plant.*, 45, 170, 1979.
34. **Sugawara, Y. and Sakai, A.**, Survival of suspension cultured sycamore cells cooled to the temperature of liquid nitrogen, *Plant Physiol.*, 54, 722, 1974.
35. **Withers, L. A.**, The freeze-preservation of synchronously dividing cultured cells of *Acer pseudoplatanus*, *Cryobiology*, 15, 87, 1978.
36. **Chen, T. H. H., Kartha, K. K., Constabel, F., and Gusta, L. V.**, Freezing characteristics of cultured *Catharanthus roseus* (L.) G. Don cells treated with dimethylsulfoxide and sorbitol in relation to cryopreservation, *Plant Physiol.*, 75, 720, 1984.
37. **Chen, T. H. H., Kartha, K. K., and Gusta, L. V.**, Cryopreservation of wheat suspension culture and regenerable callus, *Plant Cell Tissue Organ Cult.*, 4, 101, 1985.
38. **Kartha, K. K., Leung, N. L., Gaudet-LaPrairie, P., and Constabel, F.**, Cryopreservation of periwinkle *Cathranthus roseus* cells cultured *in vitro*, *Plant Cell Repts.*, 1, 135, 1982.
39. **Maddox, A. D., Gonsalves, F. G., and Shields, R.**, Successful cryopreservation of suspension cultues of three *Nicotiana* species at the temperature of liquid nitrogen, *Plant Sci. Lett.*, 28, 157, 1983.
40. **Pritchard, H. W., Grout, B. W. W., and Short, K. C.**, Osmotic stress as a pregrowth procedure for cryopreservation. III. Cryobiology of sycamore and soybean cell suspensions, *Ann. Bot.*, 57, 379, 1986
41. **Pritchard, H. W., Grout, B. W. W., and Short, K. C.**, Osmotic stress as a pregrowth procedure for cryopreservation. I. Growth and ultrastructure of sycamore and soybean cell suspensions, *Ann. Bot.*, 57, 41, 1986.
42. **Pritchard, H. W., Grout, B. W. W., and Short, K. C.**, Osmotic stress as a pregrowth procedure for cryopreservation. II. Water relations and metabolic state of sycamore and soybean cell suspensions, *Ann. Bot.*, 57, 372, 1986.
43. **Tao, D. and Li, P. H.**, classification of plant cell cryoprotectants. *J. Theor. Biol.*, 123, 305, 1986.
44. **Ulrich, J. M., Finkle, B., Moore, P. H., and Ginoza, H.**, Effect of a mixture of cryoprotectants in attaining liquid nitrogen survival of callus cultures of a tropical plant, *Cryobiology*, 16, 550, 1979.
45. **Withers, L. A.** The development of cryopreservation techniques for plant cell, tissue and organ cultures, in proc. 5th Int. Congr. Plant Tissue Cell Cult. Fujiwara, Ed., Tokyo, 1982, 793.
46. **Withers, L. A. and King, P. J.**, Proline - A novel cryoprotectant for the freeze-preservation of *Zea mays*, *Plant Physiol.*, 64, 675, 1979.
47. **Withers, L. A. and King, P. J.**, a simple freezing unit and routine cryopreservation method for plant cell cultures, *Cryo. Lett.*, 1, 213, 1980.
48. **Weber, G., Roth, E. J., and Schweiger, H. G.**, Storage of cell suspensions and protoplasts of *Glycine max* (L.) Merr, *Brassica napus* (L.), *Datura innoxia* (Mill.) and *Daucus carota* (L.) by freezing, *Z. Pflanzenphysiol.*, 109, 29, 1983.
49. **Rall, W. F. and Fahy, G.M.**, Ice-free cryopreservatrion of mouse embryos at −196°C by vitrification, *Nature (London)*, 313, 573, 1985.
50. **Bajaj, Y. P. S.**, Tuberization in potato plants regenerated from freeze-preserved meristems, *Crop Improv.*, 5, 137, 1978.
51. **Bajaj, Y. P. S.**, Production of normal seeds from plants regenerated from the meristems of *Arachis hypogaea* and *Cicer arietinum* cryopreserved for 20 months, *Euphytica*, 32, 425, 1983
52. **Mazur, P.**, Cryobiology: the freezing of biological systems, *Science*, 168, 939, 1970.
53. **Mazur, P.**, Freezing injury to plants, *Annu. Rev. Plant Physiol.*, 20, 419, 1969.
54. **Finkle, B. J. and Ulrich, J.**, Effects of cryoprotectants in combination on the survival of frozen sugarcane cells, *Plant Physiol.*, 63, 598, 1979.
55. **Finkle, B. J. and Ulrich, J. M.**, Cryoprotectant removal as a factor in the survival of frozen rice and sugarcane cells, *Cryobiology*, 19, 329, 1982.
56. **Uemura, M. and Sakai, A**, Survival of carnation *(Dianthus caryophyllus* L.) shoot apices frozen to the temperature of liquid nitrogen, *Plant Cell Physiol.*, 21, 85, 1980.
57. **Towill, L.E.**, *Solanum etuberosum:* a model system for studying the cryobiology of shoot-tips in the tuber-bearing *Solanum* species, *Plant Sci. Lett.*, 20, 315, 1981.
58. **Grout, B. W. W. and Henshaw, G. G.**, Freeze-preservation of potato shoot-tip cultures, *Ann. Bot.*, 42, 1227, 1978.
59. **Watanabe, K., Mitsuda, H., and Yamada, Y.**, Retention of metabolic and differentiation potentials for green *Lavandula vera* callus after freeze-preservation, *Plant Cell Physiol.*, 24, 119, 1983.

60. **Finhkle, B.J., Ulrich, J. M., Tisserat, B., and Rains, D. W.,** Regeneration of date palm and alfalfa plants after freezing callus tissue to $-196°C$ in a combination of cryoprotective agents, *Cryobiology,* 17, 625, 1980.
61. **Tisserat, B., Ulrich, J. M., and Finkle, B.J.,** Cryogenic preservation and regeneration of date palm tissue, *HortScience,* 16, 47, 1981.
62. **Ulrich, J. M., Mickler, R. A., Finkle, B. J., and Karnosky, D. F.,** Survival and regeneration of American elm callus cultures after being frozen in liquid nitrogen, *Can. J. For. Res.,* 14, 750, 1984.
63. **Butenko, R. G., Popov, A. S., Volkova, L. A., Chernyak, N. D., and Nosov, A. M.,** Recovery of cell cultures and their biosynthetic capacity after storage of *Dioscorea deltoidea* and *Panax ginseng* cells in liquid nitrogen, *Plant Sci. Lett.* 33, 285, 1984.
64. **Hauptmann, R. M. and Widholm, J. M.,** Cryostorage of cloned amino acid analog-resistant carrot and tobacco suspension cultures, *Plant Physiol.* 70, 30, 1982.
65. **Chen, T. H. H. and Kartha, K. K.,** unpublished results.

Chapter 11

PROPERTIES OF PEACH FLOWER BUDS WHICH FACILITATE SUPERCOOLING

Edward N. Ashworth

The dormant flower bud primordia of peach[1] and a number of other woody plant species[2-8] have been observed to supercool to low temperatures prior to freezing. In these tissues, winter survival depended on avoiding the injurious effects of ice formation. When a critical temperature was reached, the sudden freezing of a fraction of supercooled water occured. This freezing event corresponded to the death of the flower primordium.[1,3,4,8] Although supercooling is a common feature of a number of commercially important species, a clear understanding of the features which facilitate the supercooling of water in plant tissues is not known. In addition, it is not known what factors limit the extent of supercooling and the cold hardiness of these tissues. The purpose of this chapter will be to discuss the various properties of peach flower buds which appear to facilitate supercooling and suggest further avenues of research.

Water in dormant peach flower buds has been observed to freeze as two distinct components (Figure 1). The first freezing event occured a few degrees below 0°C and the second at much lower temperatures, generally between -15 and -25°C. The first freezing event, often referred to as the high temperature exotherm, corresponded to the freezing of water in the bud axis and scales and was not injurious.[9,10] The second or low temperature exotherm corresponded to the freezing of a supercooled fraction of water and was closely correlated to the temperature at which the bud primordia were killed.[10] Bud primordia apparently lack any tolerance for ice formation and once initiated, ice spread throughout and the tissue was killed as a unit.[10]

For water within a flower bud to supercool, three criteria must be met. First, the primordium must be free of heterogenous ice nuclei. Second, the spread of ice from adjacent tissues into the primordium must be prevented and third, the complete redistribution of liquid water in the primordium to ice in adjacent tissues must be prevented. In order for plant tissues to meet these criteria, some researchers have proposed that a physical or thermodynamic barrier must exist to prevent the direct growth of ice into the primordium from adjacent tissues, and the redistribution of liquid water to ice in adjacent tissues.

The observations by George and co-workers[4] that both living and dead azalea florets supercooled, led them to conclude that structural features were important in supercooling, and that the nucleation barrier was organized at the tissue level. Subsequent studies with peach have supported this proposal. If dormant buds were excised so as to remove too much of the subtending tissue[10] or if buds were pierced with a fine disecting needle,[9] subsequent supercooling was prevented. Presumably these treatments disrupted tissue structural features and provided a pathway for the propagation of ice into the primordium. Although tissue structure appeared to be important, no structural barriers were apparent at the light microscope level.[9] Cells within the dormant bud primordium were smaller than those in the subtending axis tissue. In addition, these cells were densely packed and little intercellular space was observed. However, despite these anatomical differences, there was not a marked distinction in cell size and intercellular spacing in the zone where the bud axis and primordium join.[9]

The viability of peach bud tissue had a pronounced effect on supercooling. Buds killed by either heat treatment or by rapid freezing on dry ice did not exhibit supercooling.[9] However, isolated primordia exposed to these same treatments would still supercool to low temperatures prior to freezing.[9] Additional experiments demonstrated that if freezing pre-

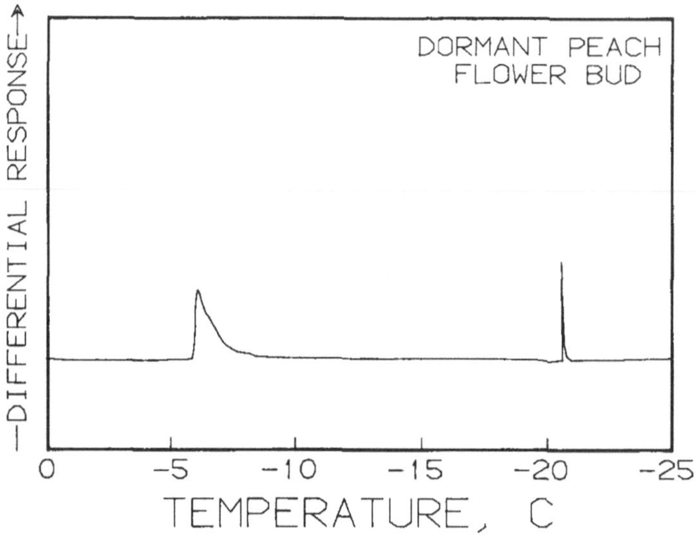

FIGURE 1. Differential thermal analysis of a dormant peach flower bud cooled at 5°C/hr.

treatments killed only the primordia of intact buds and not the subtending tissues, water within the dead primordia would still supercool. It was only treatments which killed the entire bud which prevented subsequent supercooling.[9] These results noted an interesting phenomenon where the properties of one tissue, the bud axis, influenced the freezing behavior of water in another, the bud primordium. These experiments also demonstrated that tissue structure alone, does not account for the supercooling phenomena.

One aspect of bud anatomy and morphology which appeared to influence freezing behavior was vascular development. Dormant peach bud primordia were not connected to the remainder of the plant by functional xylem vessels.[9] Although vascular strands were observed throughout the dormant bud, these tissues had not fully differentiated. Cells within the vascular strands were elongated, contained densely stained cytoplasm and lacked lignified secondary wall thickenings.[9,11] These cells were identified as procambial cells. It was hypothesized that the lack of vascular development was an important feature of buds which exhibit supercooling.[9] If xylem vessels did connect the dormant primordium to the subtending tissues, the vessels would provide a conduit for the spread of ice into the primordium, and thereby prevent supercooling. Further studies demonstrated that a lack of xylem vessel continuity between the primordium and subtending tissue was a common feature of dormant *Prunus* flower buds which supercool. Experiments indirectly accessing vascular development by following the movement of a water soluble dye[2] and anatomical studies on *Prunus* species which had varying floral morphologies[11] both found a correlation between the lack of xylem vessel continuity and the capacity to supercool. In addition, in each of the species examined, the capacity to supercool was lost during the same time interval that xylem vessel continuity was established. This suggested that the resumption of vascular differentiation in the spring may be involved in the deacclimation of buds. Once vascular continuity was established, the vessels would provide a conduit for the spread of ice, and presumably the capacity to supercool would be lost.[9,11]

The rate at which dormant buds of peach[9,12] and other species[3,4,13,15] were cooled had a pronounced effect on whether water in the primordia would supercool and the temperature of low temperature exotherm. Peach buds cooled at either 20 or 120°C/hr did not supercool while those cooled 1 or 5°C/hr did.[9,12] A similar effect has been reported with other species.[34,13,15] In contrast, excised flower primordia supercooled when cooled at either 2 or

20°C/hr.[9] This observation suggested that cooling rate did not affect the freezing behavior of water in the dormant primordium per se. This was further illustrated in an experiment when dormant buds were cooled slowly to −10°C, and subsequently cooled at 20°C/hr. In this instance, the water within the primordium would supercool. Therefore, the cooling rate dependence was expressed between 0 and −10°C, the range where ice formation occurred in the bud axis and scales.[9] The combined evidence suggested that the freezing of water within the bud axis and scales influenced the freezing behavior of the bud primordium. If buds were cooled at sufficiently slow rates, then water within the axis and scale tissue would freeze in such a manner to effectively isolate the liquid water within the primordium such that supercooling would occur.

As noted earlier, ice formation within peach buds was initiated within the bud axis and scale tissues and the freezing behavior of these tissues had a pronounced effect on the freezing of water within the dormant bud primordia. An examination of frozen buds revealed that ice crystals were not uniformly distributed within these tissues. Instead, large ice crystals were present within the bud scales and axis tissue and no crystals were noted within the primordium of viable buds.[10,16] The large crystals in the bud scales occurred within the scale tissue rather then between scales.[10] The formation of these crystals appears to have split the bud scale and disrupted many of the internal cells.[17] Likewise, within the bud axis tissue, a large crystal formed in the central portion and below it evidence of numerous smaller crystals was apparent.[17] The formation of the large central ice crystal apparently disrupted the axis tissue such that following numerous freeze-thaw cycles, evidence of a cavity was apparent in unfrozen buds.[17] In both the bud axis and scales, the large size of these crystals suggested that water had been withdrawn from other portions of the bud and had crystallized out in the scales.[10,17] Ishikawa and Sakai[18] have termed these patterns of freezing, extraorgan freezing to describe the condition where there was a segregation of ice to preferred sites such as the bud scales and crown tissue and an accompanying dehydration of the bud primordium.

The segregation of ice during freezing appears to be critical for survival of the bud primordium. Bud primordia do not appear to have any tolerance for the presence of ice crystals. Therefore during freezing, water from the primordium and subtending tissue is withdrawn and crystalizes out in portions of the bud which are not sensitive.[10,14,16,18] The redistribution of water during freezing has been measured in several species.[10,12,14,15] When cooled slowly or when held at subzero temperatures for extended periods, water continued to slowly be removed from the primordium. In the case of peach[12] and some Rhododendron species,[14] the removal of water from the primordium lead to further cold hardening. In the extreme case, Sakai[19] has described a situation in very hardy conifer buds where all the freezable water was removed and the dehydrated shoot primordia were able to survive temperatures far below the homogeneous nucleation point. Sakai[19] noted that the more cold hardy conifer species had smaller shoot primordia and he speculated that the smaller size may facilitate the dehydration of the tissue during extraorgan freezing.

Although it has become clear that a redistribution of water occurs within buds during freezing, and that this is important for winter survival, it is not clear why ice formation is first initiated in the bud scales and how water migrates to the growing ice crystals, rather than ice proliferating throughout the bud tissues. It has been proposed that the bud scales may contain intrinsic ice nucleators[14,15,18] and that differences in water potential among the tissues leads to the redistribution of tissue moisture.[10,12,14,15,18]

We have observed that xylem vessels connect the woody tissue of the peach tree to the bud scales and subtending axis tissue but not the dormant bud primordium.[9,17] Ice formation occurred between −2 and −3°C in peach trees under field conditions.[20,21] Once initiated, ice could spread via the vascular system and initiate freezing within the bud axis and scales, but would not spread into the dormant primordium. As ice formed within the bud axis and scales, a vapor pressure gradient between these tissues and the primordium would be established. This would lead to a movement of water from the primordium to the growing ice crystals within the bud axis and scales. As discussed previously, this redistribution of water within buds during freezing appears to be an important step required to isolate the liquid water within the bud primordium from ice crystals in adjacent tissues.

It is still not clear how water migrates to the growing ice crystals outside the primordia. Ishikawa and Sakai[15] noted by infiltrating buds with silicone oil, that water probably does not migrate via the vapor phase between the bud scales. Instead they propose that water moves via the vascular system to the scales. This explanation seems rather unlikely in peach buds. In peach, vessels do not connect the dormant primordia to the subtending tissue and if they did, would provide an ideal conduit for the spread of ice into the tissue. Another possibility is that water migrates through the apoplast in the vapor phase. This would be analogous to the redistribution of ice in soils during frost heaving. In some soils, large ice lenses form and water vapor migrates through the soil to the growing crystals.[22] In soils, the moisture content and soil particle size are important in determining whether ice will propagate throughout the soil, or whether ice lenses will form and a redistribution of soil moisture occurs.[22] While this may be an attractive hypothesis because it requires no special barriers, it will be difficult to test and must await further research.

In summary, the capacity of dormant peach flower buds to exhibit deep supercooling appears to be a composite of numerous features. A number of structural, morphological, and physiological features have been implicated. Presumably each feature is critical, and the absence of any one would prevent supercooling. What limits the hardiness of these buds and how these buds lose hardiness in the spring is still unknown.

REFERENCES

1. **Quamme, H. A.,** An exothermic process involved in the freezing injury to flower buds of several *Prunus* species, *J. Am. Soc. Hortic. Sci.*, 99, 315, 1974.
2. **Ashworth E. N. and Rowse, D. J.,** Vascular development in dormant *Prunus* flower buds and its relationship to supercooling, *HortScience*, 17, 790, 1982.
3. **Biermann J., Stushnoff, C., and Burke, M. J.,** Differential thermal analysis and freezing injury in cold hardy blueberry flower buds, *J. Am. Soc. Hortic. Sci.*, 104, 444, 1979.
4. **George M. F., Burke, M. J., and Weiser, C. J.,** Supercooling in overwintering azalea flower buds, *Plant Physiol.*, 54, 29, 1974.
5. **Nus, J. L., Weigle, J. L., Schradle, J. J.,** Superimposed amplified exotherm differential thermal analysis system, *HortScience*, 16, 753, 1981.
6. **Pierquet, P., Stushnoff, C., and Burke, M. J.,** Low temperature exotherms in stem and bud tissues of *Vitis riparia* Mischx, *J. Am. Soc. Hortic. Sci.*, 102, 54, 1977.
7. **Rajashekar, C. and Burke, M. J.,** The occurrence of deep undercooling in the genera, *Pyrus, Prunus,* and *Rosa*: a preliminary report, in *Plant Cold Hardiness and Freezing Stress,* Li, P. H. and Sakai, A., Eds., Academic Press, New York, 1978, 213.
8. **Sakai, A.,** Deep supercooling in winter flower buds of *Cornus florida* L., *HortScience*, 14, 69, 1979.
9. **Ashworth, E. N.,** Properties of peach flower buds which facilitate supercooling, *Plant Physiol.*, 70, 1475, 1982.
10. **Quamme, H. A.,** Mechanism of supercooling in overwintering peach flower buds, *J. Am. Soc. Hortic. Sci.*, 103, 57, 1978.
11. **Ashworth, E. N.,** Xylem development in *Prunus* flower buds and the relationship to deep supercooling, *Plant Physiol.*, 74, 863, 1984.
12. **Quamme, H. A.,** Relationship of air temperature to water content and supercooling of overwintering peach flower buds, *J. Am. Soc. Hortic. Sci.*, 108, 697, 1983.
13. **Burke, M. J. and Stushnoff, C.,** *Frost Hardiness.* A discussion of the possible molecular causes of injury with particular reference to deep supercooling of water in *Stress Physiology in Crop Plants,* Mussell, H. and Stapels, R. C., Eds., John Wiley & Sons, New York, 1979, 199.

14. **Ishikawa, M. and Sakai, A.,** Freezing avoidance mechanisms by supercooling in some *Rhododendron* flower buds with reference to water relations, *Plant Cell Physiol.*, 22, 953, 1981.
15. **Ishikawa, M. and Sakai, A.,** Extraorgan freezing in winter flower buds of *Cornus officinalis* Sieb. et Zucc., *Plant Cell Environ.* 8, 333, 1985.
16. **Dorsey, M. J.,** Ice formation in fruit buds of peach, *Proc. Am. Soc. Hortic. Sci.*, 31, 22, 1934.
17. **Ashworth, E., Davis, G., and Wisniewski, M.,** Unpublished data.
18. **Ishikawa, M. and Sakai, A.,** Characteristics of freezing avoidance in comparison with freezing tolerance. A demonstration of extraorgan freezing, in *Plant Cold Hardiness and Freezing Stress,* Vol 2, Li, P. H. and Sakai, A., Eds., Academic Press, London, 1982, 325.
19. **Sakai, A.,** Freezing avoidance mechanisms of primordial shoots of conifer buds, *Plant Cell Physiol.*, 20, 1381, 1979.
20. **Ashworth, E. N. and Davis, G. A.,** Ice nucleation within peach trees, *J. Am. Soc. Hortic. Sci.*, 109, 198, 1984.
21. **Ashworth, E. N., Anderson, J. A., Davis, G. A., and Lightner, G. W.,** Ice formation in *Prunus persica* under field conditions, *J. Am. Soc. Hortic. Sci.*, 110, 322, 1985.
22. **Miller, R. D.,** Freezing phenomena in soils, in *Applications of Soil Physics,* Hillel, D., Ed., Academic Press, New York, 1980, 254.

Chapter 12

FORECASTING HARDINESS FOR WINTER COLD PROTECTION IN PEACH ORCHARDS

E. L. Proebsting, Preston K. Andrews, Anne Swindeman-Murphey, and Guy Reisenauer

Peach flower buds are susceptible to low temperature injury during the winter months in the higher latitude peach growing districts of North America. This condition is controlled only by selecting favorable orchard sites. The use of hardier cultivars and hardiness-promoting cultural systems have not solved the problem.

There are differences in the hardiness of cultivars that appear during winters which are neither cold enough to kill all the buds nor mild enough to cause no damage. Averaging five such "test" winters, the range of survival among 24 cultivars was from 88% for the hardiest to 25% for the most tender.[13] A temperature range of ~2°C results in a similar range of damage.[20]

Most commercial peach or nectarine growers in the northern districts compete on the market with fruit grown in warmer areas where winter injury to flower buds is not a problem, Unless the northern grower has a large advantage in delivering more mature fruit, it is usually necessary to grow the same cultivars originated and grown in the warmer areas. The choice of cultivar is often based on fruit type and ripening season, rather than bud hardiness. So far, breeding has not been a powerful force in reducing losses of flower buds to low winter temperatures.

Cultural practices have similarly failed to increase bud hardiness substantially. Gibberellic acid (GA_3) increases bud hardiness by up to 1°C[6,9,16] as does ethephon.[8,19] Use of these treatments is discouraged, however, as GA_3 reduces wood hardiness and ethephon reduces fruit size.

Frost control by modification of the atmosphere in the orchard has been practiced on a commercial scale for most of this century.[25] This was accomplished first by burning many small fires, later by using large fans that mix cold air at the surface with warmer air a few meters above the trees, and most recently by applying water on or under the tree. Where water is used, the source of heat is the exothermic phase change of water from liquid to solid.

These methods for modifying the atmosphere are used almost exclusively where the plant tissues being protected are in or near the frost tender state, having critical temperatures of −2 to −5°C. Citrus during midwinter at subtropical latitudes, and deciduous fruit during spring bloom at higher latitudes are examples of crops which must be protected. Since deciduous fruit may also be subject to low temperature injury during midwinter and since many growers have capital invested in frost protection equipment already, it is logical to consider using the equipment during midwinter periods as well.[17] Innovative central Washington peach and nectarine growers have done this now for 10 to 15 years.

There are several factors that make dormant season protection of deciduous fruit more difficult than protecting during the cold tender state, however. One is the very low temperatures (−20 to −25°C) that occur. Such low temperatures are not only uncomfortable to work in, but may be hazardous to those operating the equipment. Special attention is required to start and run equipment at such low temperatures. One must also know whether inversions will be present because heaters and wind machines depend on strong inversions to be most effective. Methods using water are considered to be inoperable at such low temperatures.

Perhaps the most interesting problem in midwinter cold protection for deciduous fruit trees is the rapid fluctuation in their critical temperatures in response to ambient temperature. Flower buds are damaged more severely following mild temperatures than following prolonged cold weather. Early measurements of cold hardiness in peach flower buds using laboratory freezing techniques document such fluctuations.[4,5,7,12] Subsequent detailed studies show some order in the manner of these fluctuations.[14,15,17]

In brief, the buds have a minumum hardiness level (MHL) that is very stable and is associated with the stage of development of the flower bud.[15] Buds do not become appreciably less hardy than the MHL regardless of how warm the weather is during midwinter.[15,17] The MHL is believed to be constant during late fall and winter, then rising in late winter and early spring as the early stages of development toward anthesis begin.[17] Typical MHL values are −21°C for peach and −22°C for sweet cherry during midwinter. Large increases in hardiness well below the MHL occur when the buds are frozen. Buds increase in hardiness at about 1°C per day for peaches and 2°C per day for cherries.[15,17] Hardiness of peach buds may reach −26 to −28°C following prolonged low temperatures while sweet cherry bud hardiness has reached −35°C. Temperatures mild enough to thaw the buds result in a rapid rise back to the MHL, usually within 1 day.

Understanding that dormant peach and cherry flower buds respond to ambient temperatures in a systematic fashion, the development of a standardized method for testing makes it possible to provide estimates of critical temperatures during the winter. These estimated critical temperatures can then be used to determine when to begin cold protection.[18] A few peach and nectarine growers operate heaters and wind machines successfully at temperatures of −20°C and below using this information.

Measurement of air temperatures 16 m above the ground show that inversions are common during the coldest midwinter nights in south central Washington state. Very strong inversions develop over snow on calm, clear nights[10]

The National Weather Service alerts the Washington State University Research Center at Prosser of impending dangerous low temperatures. This usually allows sufficient time to measure the current bud hardiness, which is then reported as the temperature required to kill 50% of the flower buds (T_{50}). This procedure normally is run over 3 days but can be done in one long day to provide the most current estimate of the T_{50}. At the very best it requires a forecast of nearly 24 hr to estimate the critical temperature for the next morning.

A faster method to estimate critical temperatures developed from the discovery that flower buds of *Azalea*[11] and *Prunus* species[22] survive low temperature injury by deep supercooling. When deep supercooled buds freeze, the event is nearly instantaneous, always fatal, and accompanied by an exotherm which is readily measurable with differential thermal analysis (DTA). By using a sample size of 20 to 50 buds[1,21] and observing exotherms as they occur with decreasing temperature, it is simple to calculate the temperature required to produce low temperature exotherms from 50% of the floral primordia (LTE_{50}). The rate of temperature fall can be increased from our normal 1°C/hr to 5°C/hr to permit an estimate of the present day's T_{50} and allow more time for one to decide which cold protection methods will be necessary during the night and early morning hours.

As has been shown, estimation of the T_{50} requires a forecast because the T_{50} changes significantly with time while the buds are dormant. The length of the forecast has been reduced by adopting DTA to estimate the T_{50}. Nevertheless, because the meteorological and horticultural factors are constantly changing, a forecast of the T_{50} to match the period of the longest weather forecast would be useful.

The development of quantitative relationships between temperature and LTE_{50} for populations of peach or cherry flower buds[2] permitted the development of a computer model that estimates LTE_{50}.[3]

The model utilizes several long standing concepts. One of these is the use of the accumulation of chill units to represent time. Although this assumption has not been verified, we assume that a physiological representation of time as measured by chill unit accumulation is more appropriate than calendar date. A chill unit accumulation model developed from data collected at the orchard used for the hardiness studies[24] is incorporated into the computer model.

The start of chill unit accumulation each autumn corresponds to time zero in the model. In central Washington, chill units usually begin to accumulate in early October when cherry flower buds have acquired the capacity to deep supercool. Peaches and apricots develop that capability about 2 weeks later. Flower primordia that deep supercool in October freeze at temperatures below $-15°C$, which is much colder than temperatures normally encountered at that time of year. Until mid November, the vegetative tissues are more susceptible to freezing injury than are the floral tissues.

Also assumed is an MHL, above which the LTE_{50} will not rise regardless of temperatures occurring during the dormant season. The MHL is stable until bud development begins in early spring and the primordia lose their ability to deep supercool.

The model quantifies the relationship that shows the T_{50} lowering slowly when the buds are frozen and rising rapidly to the MHL when they thaw.[15,17,22] It also deals with the responses that occur when temperatures persist near the freezing point of the buds, i.e., about $-2°C$.

The decision pathway of the model is based on a measured LTE_{50}, a calculated LTE_{50} and hourly air temperatures. Air temperatures may be measured and entered hourly, averaged from more frequent measurements, or interpolated from less frequent measurements. As presently programmed, the model approximates hourly air temperatures from a sine curve which is generated from daily measured maximum and minimum temperatures. The generated hourly temperatures are then used to calculate hourly LTE_{50} s. Each calculated LTE_{50} then becomes the base from which the next hour's LTE_{50} is calculated. Chill unit accumulation also is calculated from the temperature data base in a subprogram.[24]

The changes that occur near the freezing point of the buds are critical to the accuracy of the model. Thawing is particularly critical because the rates of change are greater than an order of magnitude faster than changes during freezing. From the limited data available, a series of equations was developed that incorporate three principal considerations: (1) the temperature of thawing, (2) the duration of thawing, and (3) the current hardiness of the flower buds. The equations reflect slow thawing near the freezing point of the buds at -1 and $-2°C$.

Observations of the hardiness behavior of dormant peach and cherry buds have shown that there are limits to the LTE_{50}. The MHL, described previously, usually ranges from -18 to $-22°C$. The lowest measured LTE_{50} for sweet cherries is near $-35°C$.[2] The computed values of the model are thus kept within these limits.

The model was validated during the 2 years which followed the development of the model. The model estimated the LTE_{50} with reasonable accuracy during these years, but discrepancies were also revealed.

The most significant discrepancy was that the MHL, which was assumed to be constant during the fall and winter, changes as the season advances. Following the onset of chill unit accumulation, the MHL decreases at an average rate of 0.56°C per 100 chill units accumulated until the buds are fully dormant and can no longer deharden with increasing temperature. The MHL increases only after chilling is completed and bud development begins; however, the model does not yet include that part of the season.

Other discrepancies have not been as clearly defined, but two types of errors are appearing that warrant further research. First, the maximum hardiness level that the buds reach after prolonged freezing is not constant, either among years or within years. The factors controlling maximum hardiness, however, are not known. The second type of error arises from the sensitivity of the flower buds to freezing and thawing. The errors in measuring or estimating air temperature, the accuracy with which air temperature reflects bud temperature, and the assumptions about the rate of change in hardiness near the freezing point all become very important when temperatures remain close to the freezing point of the buds.

As presently programmed, the model computes chill units and LTE_{50} values hourly from the hourly temperatures approximated from the sine curve. Measured hourly temperatures, if available, may be entered instead. The maximum hardiness level, the initial MHL (at the beginning of chill unit accumulation), the rate of decrease in the MHL during hardening off, the dormant season MHL, and the freezing point of the buds may be varied in the model. It appears to be possible to experimentally determine each of these values early in the season and program the model for the specific conditions of that year.

In conclusion, critical temperatures for flower bud injury vary widely and rapidly in dormant flower buds. It is an important concept to understand if midwinter protection is to be successful. The model described here is available to extension personnel, growers, and consultants; however, it has only been tested in central Washington. It provides estimates of critical temperatures to supplement measured LTE_{50} values. It also can be used to predict LTE_{50} s by utilizing forecasted temperatures obtained from the National Weather Service. Originally developed for sweet cherries, it can be readily adapted to peach and nectarine, where the need is greatest. Adaptation to the estimation of *Vinifera* grape bud hardiness also seems to be feasible, and its applicability to certain aspects of the hardiness of the vegetative tissues of *Prunus* trees will be investigated.

REFERENCES

1. **Andrews, P. K., Proebsting, E. L., and Campbell, G. S.,** An exotherm sensor for measuring the cold hardiness of deep supercooled flower buds by differential thermal analysis, *HortScience* 18, 77, 1983.
2. **Andrews, P. K. and Proebsting, E. L., Jr.,** Effects of temperature on the deep supercooling characteristics of dormant and deacclimating sweet cherry *(Prunus avium* L.) flower buds, *J. Am. Soc. Hortic. Sci.*, 112, 320, 1987.
3. **Andrews, P. K., Proebsting, E. L., and Lee, G. S.,** A conceptual model of changes in deep supercooling of dormant sweet cherry *(Prunus avium)* flower buds, *J. Am. Soc. Hortic. Sci.*, (in press), 1987.
4. **Chandler, W. H.,** Cold resistance in horticultural plants: a review, *Proc. Am. Soc. Hortic. Sci.*, 64, 552, 1954.
5. **Chaplin, C. E.,** Some artificial freezing tests of peach fruit buds, *Proc. Am. Soc. Hortic. Sci.*, 52, 121, 1948.
6. **Corgan, J. N. and Widmoyer, R. B.,** The effects of gibberellic acid on flower differentiation, date of bloom, and flower hardiness of peach, *J. Am. Soc. Hortic. Sci.*, 96, 54, 1970.
7. **Cullinan, F. P. and Weinberger, J. H.,** Studies on the resistance of peach fruit buds to injury at low temperature, *Proc. Am. Soc. Hortic. Sci.*, 32, 244, 1934.
8. **Dennis, F. G., Jr.,** Trials of ethephon and other growth regulators for delaying bloom in tree fruits, *J. Am. Soc. Hortic. Sci.*, 101, 241, 1976.
9. **Edgerton, L. J.,** Some effects of gibberellin and growth retardants on bud developement and cold hardiness of peach, *Proc. Am. Soc. Hortic. Sci.*, 88, 197, 1966.
10. **Geiger, R.,** *The Climate Near the Ground*, Harvard University Press, Camgridge, Mass, 1966.
11. **Graham, P. and Mullin, R.,** The determination of lethal freezing temperatures of deciduous azalea by a freezing curve method, *J. Am. Soc. Hortic. Sci.*, 100, 3, 1975.
12. **Meader, E. M. and Blake, M. A.,** Seasonal trend of fruit bud hardiness in peaches, *Proc. Am. Soc. Hortic. Sci.*, 43, 91, 1943.
13. **Mowery, J. B.,** Inheritance of cold hardiness of dormant peach flower buds, *Proc. Am. Soc. Hortic. Sci.*, 85, 128, 1964.
14. **Proebsting, E. L., Jr.,** Cold hardiness of Elberta peach fruit buds during four winters, *Proc. Am. Soc. Hortic. Sci.*, 74, 144, 1959.
15. **Proebsting, E. L., Jr.,** The role of air temperatures and bud developement in determining hardiness of dormant Elberta peach fruit buds, *Proc. Am. Soc. Hortic. Sci.*, 83, 259, 1963.

16. **Proebsting, E. L., Jr., and Mills, H. H.,** Gibberellin-induced hardiness responses in Elberta peach flower buds, *Proc. Am. Soc. Hortic. Sci.*, 85, 134, 1964.
17. **Proebsting, E. L., Jr., and Mills, H. H.,** A comparison of hardiness responses in fruit buds of "Bing" cherry and "Elberta" peach, *J. Am. Soc. Hortic. Sci.*, 97, 802, 1971.
18. **Proebsting, E. L., Jr., and Mills, H. H.,** The use of controlled freezing techniques for estimating critical temperatures for fruit buds, Wash. Agr. Exp. Sta., Circ. 548, 1971.
19. **Proebsting, E. L., Jr., and Mills, H. H.,** Ethephon increases cold hardiness of sweet cherry, *J. Am. Soc. Hortic. Sci.*, 101, 31, 1976.
20. **Proebsting, E. L., Jr., Mills, H. H., and Russell, T. S.,** A standardized temperature-survival curve for dormant Elberta peach fruit buds, *Proc. Am. Soc. Hortic, Sci.*, 89, 85, 1966.
21. **Proebsting, E. L., Jr. and Sakai, A.,** Determining LTE_{50} of peach flower buds with exotherm analysis, *HortScience*, 14, 597, 1979.
22. **Quamme, H. A.,** An exothermic process involved in the freezing injury to flower buds of several *Prunus* species, *J. Am. Soc. Hortic. Sci.*, 99, 315, 1974.
23. **Quamme, H. A.,** Mechanisms of supercooling in overwintering peach flower buds, *J. Am. Soc. Hortic. Sci.*, 103, 57, 1978.
24. **Richardson, E. A., Seeley, S. D., and Walker, D. R.,** A model for estimating the completion of rest for "Redhaven" and "Elberta" peach trees, *HortScience*, 9, 331, 1974.
25. **Young, F. D.,** Frost and the prevention of frost damage, U.S. Department of Agriculture, Farmer's Bull., 1588, 1929.

Responses of Chilling-Sensitive Plants to Low Temperature Stress

Chapter 13

MEFLUIDIDE: A SYNTHETIC CHEMICAL THAT PROTECTS CORN AND RICE SEEDLINGS FROM CHILLING INJURY

P. H. Li

TABLE OF CONTENTS

I. Introduction 168

II. Mefluidide 168

III. Application and Results 168
A. Preparation of Solution 168
B. Effect of Mefluidide Applied 24 Hr Before Chilling on the Reduction of Chilling Injury 168
1. Corn Seedlings 168
2. Rice Seedlings 169
C. Effect of Intervals Between Mefluidide Treatment and Chilling Exposure on the Reduction of Chilling Injury 170
D. Post-Application Effect on Growth and Development With and Without Chilling 170
1. Corn 170
2. Rice 170

IV. Discussion 171
A. Water Relations 171
B. Membrane Permeability 172
C. Abscisic Acid 172

V. Concluding Remarks 175

References 176

I. INTRODUCTION

Based on volume of production, rice *(Oryza sativa)*, maize *(Zea mays)*, and wheat *(Triticum sativum)* rank among the top three major world food crops.[7] Two of the three, rice and maize, are sensitive to chilling temperature. Their growth and development can be drastically hindered when environment temperatures reach below 10°C, non-freezing temperatures.[10] Consequently, yield is decreased if it is not total crop failure. One way to increase crop production is by early planting, especially in the case of corn grown in one-seasonally cropping region.[13] However, early planting has been limited in many producing regions because of the risk of chilling stress. Overcoming chilling injury for yield improvement has been a challenge in agricultural research. Numerous attempts have been made, for example, in modifying crop practices to avoid chilling injury, breeding and selection for early vigor. Attempts to protect plants from chilling injury by application of chemicals have also been reported.[9,15]

In 1984, the Laboratory of Plant Hardiness first reported that mefluidide, a synthetic growth regulator,[2] was capable of protecting chilling sensitive plants such as cucumber and corn seedlings from severe chilling injury.[18] Subsequently, detailed studies of its protection to chilled corn[19] and rice[23] seedlings, mechanism of protection,[22] and post-application effect on growth and development[19,23] have been conducted. This chapter provides a summary review of the effectiveness of mefluidide on chilling protection of corn and rice seedlings.

II. MEFLUIDIDE

Mefluidide is a synthetic plant growth regulator.[2] It affects plant growth drastically. Depending on concentrations applied, it can stunt or stimulate plant growth at high or low concentrations, respectively. Embark has been its trade name. The chemical name is N-{2,4-dimethyl-⟨[(trifluoromethyl)sulfonyl]amino⟩phenyl}acetamide. It does not dissolve in water but will in organic solvents such as acetone. Its chemical structure is shown in Figure 1.

III. APPLICATION AND RESULTS

A. Preparation of Solution

Either analytic or technical grade (94%) mefluidide can be applied to corn and rice seedlings for chilling protection. It should be first dissolved in a minimum quantity, e.g., 2 to 3 mℓ of acetone, and then quickly diluted with water to an appropriate volume. A few drops of surfactant Tween-20 should be added to the mefluidide solution at a rate of 2 to 3 drops per 100 mℓ volume when foliar spray is used. Application stops when leaf tips show droplets. Plants sprayed with water including the same rate of acetone and Tween-20 serve as the control.

B. Effect of Mefluidide Applied 24 Hr Before Chilling on the Reduction of Chilling Injury

1. Corn Seedlings

Exposure of corn seedlings to chilling causes leaf necrosis and drying of the leaf margins. In general, fully expanded second and third leaves are most sensitive to chilling temperatures. On a per plant basis, 10, 25, and 60% of the total leaf area was damaged when chilled controls have been subjected to a constant temperature of 4°C for 2, 4, and 6 days, respectively.[19]

Plants treated with mefluidide had significantly reduced leaf damage. Leaves treated with 5, 10, 15, or 20 ppm mefluidide had about 10% damage after 6 days of exposure to a

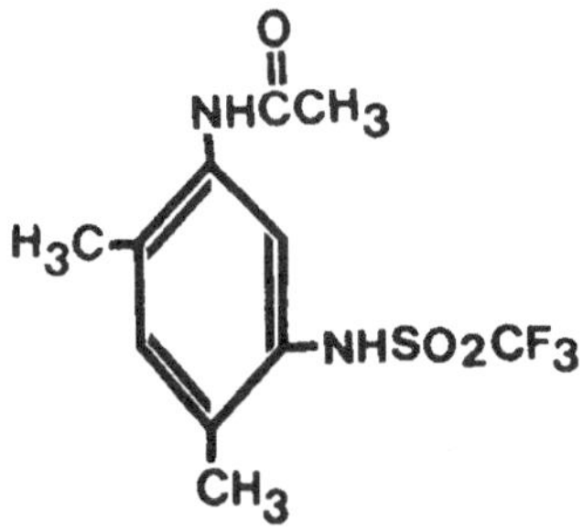

FIGURE 1. Chemical structure of mefluidide.

FIGURE 2. The appearance of corn seedlings 24 hr after a 6-day exposure to a constant temperature of 4°C. Plants were foliar sprayed with 5 ppm mefluidide (5) and water (○). The 5 ppm treated plants and controls (○) had 6 and 60% damage of the total leaf area, respectively.[19]

constant temperature of 4°C. The appearance of chilled control (0) and chilled, treated plants (5) after 6 days chilling stress is shown in Figure 2.

Plants treated with mefluidide (5, 10, 15, or 20 ppm) also maintained at a higher leaf growth rate during chilling exposure than the chilled controls.[19]

2. Rice Seedlings

Leaves of chilled rice seedlings became discolored and dried out from tip to the base of the leaf blade after 8 days exposure to a constant temperature of 6°C.[23] On a per plant basis, 65% of the total leaf area of the control was damaged. Plants treated with mefluidide at concentrations of 3, 5, 10, and 15 ppm had mean injuries of about 30%. Concentrations of 3, 5, 10, and 15 ppm mefluidide appeared to provide similar degrees of chilling protection for rice seedlings.

C. Effect of Intervals Between Mefluidide Treatment and Chilling Exposure on the Reduction of Chilling Injury

Effective chilling protection for corn seedlings was not observed unless mefluidide treatment was given in a warm regime (26°C) for 12 hr prior to chilling exposure.[22] For rice seedlings, mefluidide treatment must be held at least 6 hr before exposure to chilling temperature.[23] Protection for rice and corn seedlings from chilling injury can last for at least 72 hr after a single treatment. Protection, however, has not yet been tested beyond 72 hr in both rice and corn seedlings after a single treatment.

D. Post-Application Effect on Growth and Development With and Without Chilling

Long-term effects of chilling exposure and mefluidide treatment on corn and rice growth and development have been investigated.[19,23]

1. Corn

It is known that chilling alone can inhibit plant growth. Plants chilled and treated with mefluidide at 5, 10, and 15 ppm were significantly taller than chilled controls.[19] The differences in plant height during post-application increased with duration of growth. After 45 days growth, chilled, treated plants were not significantly different in heights from plants that were not chilled and received no mefluidide treatment.

The 20-ppm mefluidide treatment inhibits plant growth. Growth restriction with high mefluidide concentrations was accompanied by necrosis of leaf tips and margins and by plant distortion. Leaves fail to unroll, giving a "buggy whip" appearance. This phenomenon has been observed in soybeans *(Glycine max)* and red rice *(O. sativa)*.[14]

Plants that were treated with 5, 10, or 15 ppm mefluidide and subsequently chilled at 4°C for 6 days had significantly greater fresh and dry weight than chilled controls.[19] Weights of unchilled and chilled-treated plants were not significant different after 55 days growth in the greenhouse. Plants chilled and treated with 5 or 10 ppm mefluidide flowered (when silks became a full reddish-purple color) 5 to 6 days earlier than those of chilled controls, which reached flowering 86 days after planting. Plants with no treatment and no chilling flowered 72 days after planting. Although 15 ppm mefluidide-treated plants showed chilling protection and had similar plant heights and weights as 5- and 10-ppm treated plants, it appeared that 15 ppm treatment may delay flowering.

Depending on concentrations applied, melfuidide drastically affects plant growth and development. For corn seedlings, the optimum foliar spray concentration for a significant chilling protection without adverse post-application effect on growth and development was between 5 and 10 ppm (mg/liter).

2. Rice

Plant growth was investigated at nonchilling regime after plants were treated with different concentrations of mefluidide.[23] It was observed that growth was more retarded with increasing concentration of applied mefluidide. Plants treated with 5 or 10 ppm, however, did not differ in heights from the nonchilled controls. As a matter of fact, 5 ppm application may stimulate plant growth.

Plants treated with mefluidide at concentrations of 3, 5, and 10 ppm and chilled at a constant temperature of 6°C for 8 days recovered rapidly from chilling injury and grew faster than chilled controls. Plants treated with 15 ppm mefluidide showed no difference in heights from unsprayed controls after 8 days exposure to a constant temperature of 6°C. Heading date is one of the common parameters often used in estimating chilling injury in rice seedlings.[1] Chilled plants treated with 5, 10, and 15 ppm mefluidide, headed 10, 8, and 5 days earlier, respectively, than did the chilled controls (Figure 3).

It is concluded that mefluidide protects rice seedlings from chilling injury. The optimum

FIGURE 3. Rice heading dates and the appearance of plants after chilling and mefluidide treatment. June 8 and June 18, 1986, were the initial heading dates for chilled controls (CK), and chilled, 5 ppm mefluidide-treated plants, respectively. The picture was taken on June 24, 1986.

concentrations for foliar spray, without adverse growth effects, are between 3 and 10 ppm (mg/liter).

DISCUSSION

A. Water Relations

Corn plants treated with mefluidide showed no significant changes in relative water content (RWC) during and after chilling, whereas the control continuously decreased in RWC (Figure 4).[22] During a 6-day exposure to a constant temperature of 4°C, mefluidide-treated corn plants and control had a more or less constant level of water potential (between −1 and −3 bars). When plants were removed from chilling to a warm regime, water potential (WP) of the control declined as compared with chilled and mefluidide-treated plants in which decrease in WP was insignificant.[22] During chilling, plants lose water.[21] It has been suggested that abscisic acid (ABA) may aid plants in preserving better water balance and in tolerating more chilling stress.[15,16] It appears that the better water balance in chilled, mefluidide treated corn plants may result from an imbalance of endogenous plant growth regulators.

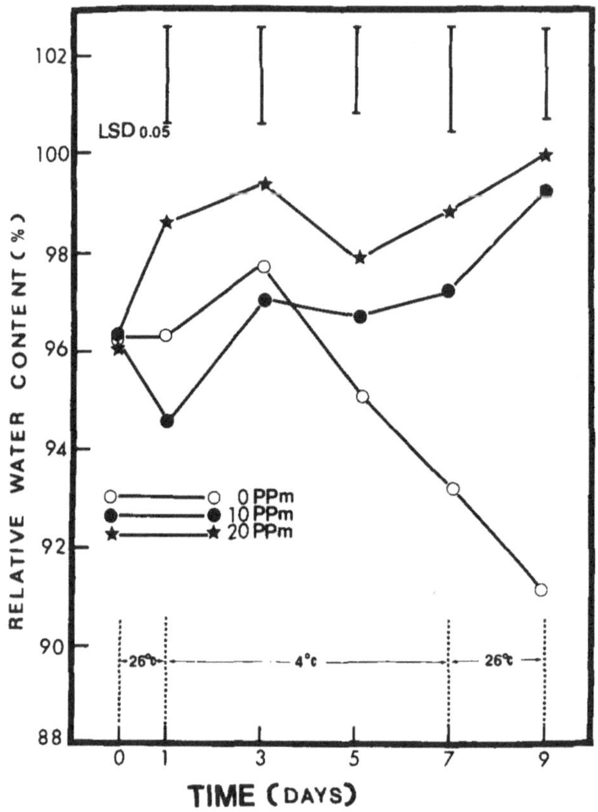

FIGURE 4. Leaf relative water contents (RWC) of chilled control and chilled, mefluidide-treated corn plants before, during and after 4°C chilling.[22]

B. Membrane Permeability

It is generally acceptable that plasma membrane is the primary site in sensing chilling injury.[10,11] Chilling-induced increase in electrolyte leakage has been used as a measurement of the integrity of plasma membrane. As the degree of chilling injury increased, so did electrolyte leakage. In corn seedlings, mefluidide treated plants showed no changes in electrolyte leakage during chilling, whereas chilled controls had an increase in conductivity from 10 to 40% (Figure 5).[19] The difference in conductivities between chilled controls and chilled mefluidide-treated plants were highly significant at 2, 4, or 6 days exposure to a constant temperature of 4°C.[19] The same is true in rice seedlings.[23] When seedlings were exposed ot 6°C, chilling-induced electrolyte leakages also increased. For example, during a 7-day chilling exposure, unsprayed chilled-control increased in conductivity from a 30% level to greater than 70%. The conductivity of chilled treated seedlings remained more or less at the 30% level.[23] Plants not chilled, but sprayed with mefluidide remained at a 5% level.

Although the mechanism by which corn and rice seedlings can be protected from chilling injury with mefluidide treatment is not understood, mefluidide appears to affect plasma membrane permeability or alter the properties of the plasma membrane. St. John and Hilton[17] demonstrated that perfluidon, which is chemically related to mefluidide, affects the synthesis of neutral and poplar lipids known to be the basic constituents of the plasma membrane.

C. Abscisic Acid

There is evidence that endogenous ABA increase is interrelated with increase in plant

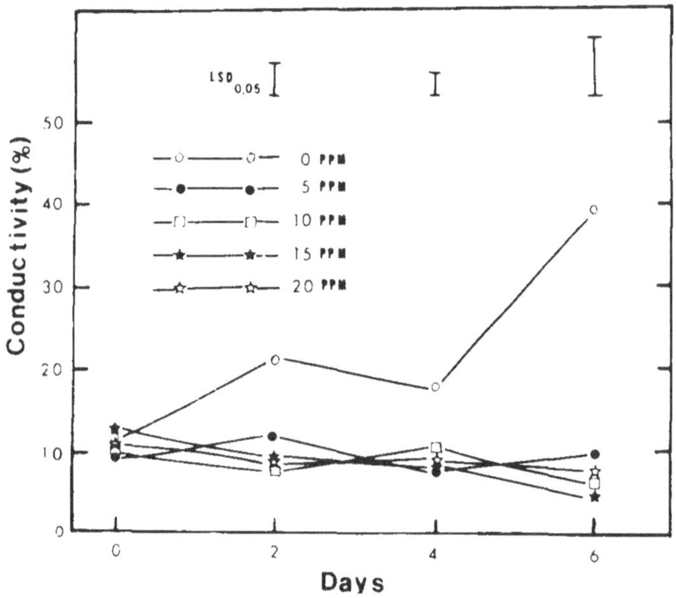

FIGURE 5. Effects of mefluidide treatments on electrolyte leakages of corn leaf tissues during 6 days exposure to constant temperature of 4°C.[19]

cold hardiness.[4] However, the increase in ABA content is not necessarily parallel with the increase in cold hardiness. In *Solanum commersonii* potato, for instance, ABA content increases during low temperature exposure, occurring prior to an increase in cold hardiness, and the increase is species specific.[4] Also, Rikin and Richmond[15] and Rikin et al., [16], have demonstrated that exogenously applied ABA ameliorated cucumber and cotton seedlings from chilling injury. Nevertheless, it is fairly certain that ABA plays a vital role in the induction of plant cold hardiness.

Change in ABA content of corn seedlings, before, during, and after chilling (4°C), treated with mefluidide at concentrations of 10 and 20 ppm, is shown in Figure 6. Immediately after mefluidide treatment, there was an increase in ABA content, even though plants were grown in a temperature of 26°C regime.[22] Two days after chilling (4°C), ABA content in the control increased 50%, and reached a level similar to the chilled mefluidide-treated plants. Thereafter, both chilled, treated plants and chilled controls maintained a similar and constant level of ABA. When plants were removed from chilling and placed at 26°C, ABA content significantly increased again in both treated and control plants. The control, which had the greatest chilling damage (60% drying of leaf to total leaf area) after 6 days chilling, showed the highest increase in ABA content.

Mefluidide seems capable of triggering an increase in ABA content of corn seedlings with the absence of chilling. The higher the concentration applied, the greater the amount of ABA increase. Chilling alone causes an increase in ABA content.[22] It is known that chilling results in water stress which in turn induces an increase in ABA content.[6] However, in corn seedlings, the increase in ABA content during the first 2 days of chilling is likely influenced by the low temperature per se rather than by the low-temperature induced loss of water, which occurred 2 days after exposure to 4°C (Figure 4). It appears that the increase in ABA content during chilling, even occurring at the early chilling stage, was not correlated to corn chilling resistance, because control plants showed more than 60% leaf damage after a 6-day chilling exposure.

The timing of elevated ABA content (before chilling) appears to be more important for chilling protection than the high level of ABA that occurs during chilling. Rikin, et al.[16]

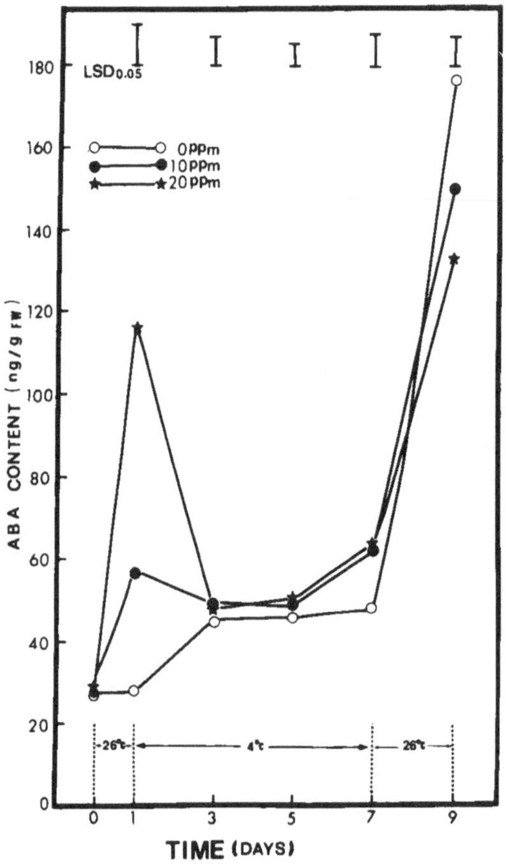

FIGURE 6. Changes in endogenous ABA content in corn leaves of controls (○) and 10 and 20 ppm mefluidide-treated plants (●, *) before, during and after 4°C chilling.[22]

have shown that ABA applied at the inception of chilling was ineffective in preventing chilling injury, and for an effective protection, application must be held at least 3 hr prior to chilling exposure. The same is true for mefluidide-induced chilling protection for corn and rice seedlings. Mefluidide treatment must be given in a warm regime (25 to 26°C) for at least 12 and 6 hr for corn and rice, respectively, prior to chilling exposure in order to have an effective protection.

It is suggested that an increase in endogenous ABA content, as triggered by mefluidide before chilling,[22] would be a necessary step in activating the protection system that enables plants to minimize the injury during low temperature exposure. Then an obvious question is, would direct ABA treatment be effective to protect corn seedings? ABA is capable of preventing plant tissues from chilling injury *in vitro*.[15,16] Unfortunately, spraying corn seedlings with ABA does not work.

In order to provide additional evidence to support the hypothesis that mefluidide triggers an ABA increase which activates the protection system, fluridone experiments were conducted in corn seedlings. It has been reported that fluridone, 1-methyl-3-phenyl-5-[3-(trifluoromethyl)phenyl]-4(1H)pyridinone, decreases endogenous ABA content in corn plants[12] by inhibiting carotenoid synthesis.[3] Table 1 shows the capability of fluridone in inhibiting the mefluidide-induced chilling protection of corn seedlings. Plants treated with mefluidide alone had about 30% chilling damage as compared to 70% for the chilled control. Leaves

Table 1
INHIBITION EFFECT OF FLURIODONE ON MEFLUIDIDE-INDUCED CHILLING PROTECTION IN CORN LEAVES

Treatments[a] %damage of total leaf area				
Control	Fluridone	Mef + Flu	Mef to Flu[b]	Mefluidide
71.1[a]	70.0[a]	65.1[ab]	49.6[b]	31.1[c]

Note: Measurements of injured leaf area were carried out from plants that were exposed to a constant temperature of 4°C for 15 days followed by a 2-day recovery. LSD 0.05

[a] Treatments: 5 mg/ℓ mefluidide (Mef), 25 mg/ℓ fluridone (Flu).
[b] Plants were treated first with mefluidide and then fluridone 24 hr after.

that were treated with either fluridone alone or a combination of mefluidide and fluridone, showed chilling damage and the degree of damage was similar to the control. Interestingly, plants that were treated with mefluidide first and fluridone 24 hr after, were chill protected. During subsequent growth observation, only those plants that were treated either with mefluidide alone or mefluidide first and then fluridone survived. Applying fluridone 24 hr after mefluidide treatment was deliberately designed. It was hypothesized that a high level of ABA content should have occurred 24 hr after mefluidide treatment as previously observed.[22] Because the high level of ABA content already present in the plant system, and thus an activation of the protection system had been initiated within 24 hr after mefluidide treatment, fluridone applied 24 hr after mefluidide treatment would have not theoretically affected the mefluidide-induced chilling protection, if the protection system is triggered by an increase in ABA content.

The inhibition of mefluidide-induced chilling protection when plants were treated with a combination of mefluidide and fluidone, and the protection observed from those plants that were first treated with mefluidide and then fluridone 24 hr after (Table 1), support the hypothesis that (1) the mechanism of mefluidide-induced chilling protection appears to be due to mefluidide's ability to trigger an increase in ABA content when plants are grown in a warm environment, and (2) the increase in ABA content then activates the protection system which in turn enables plants to minimize the chilling injury. However, ABA content in fluridone-treated plants has not yet been determined. It has been demonstrated that ABA is able to induce the synthesis of certain protein species which may be involved in the induction of plant cold hardiness.[4,20]

V. CONCLUDING REMARKS

Mefluidide protects corn and rice seedlings from chilling injury in the controlled environment. The optimum concentrations, without an adverse growth and development effect after application, are between 3 and 10 ppm (mg/ℓ). For an effective protection, mefluidide treatment must be held at least 6 and 12 hr for rice and corn, respectively, prior to exposure to chilling. Protection has been observbed 72 hr after a single application. Chilled, mefluidide-treated rice or corn plants headed or flowered about 5 to 7 days earlier, respectively, than chilled, non treated controls.

It is not known whether mefluidide will be effective in terms of improving crop productivity after chilling stress in the field. It is economically feasible for large scale evaluation and worth testing. In addition, this chemical appears to be a powerful tool in elucidating the mechanisms of chilling injury and resistance.

REFERENCES

1. **Adair, C.R.,** Testing rice seedlings for cold water tolerance, *Crop Sci.*, 8, 264, 1968.
2. Agricultural Production/3M. Technical Data Bulletin, 3M Company, St. Paul, Minnesota, 1980.
3. **Bartels, P. G. and Watson, C. W.,** Inhibition of cartenoid synthesis by fluridone and norfurazon, *Weed Sci.*, 2, 198, 1978.
4. **Chen, H. H., Li, P. H., and Brenner, M. L.** Involvement of abscisic acid in potato acclimation, *Plant Physiol.*, 71, 362, 1983.
5. **Cummins, W. R., Kende, H., and Raschke, K.,** Specificity and reversibility of the rapid stomatal response to abscisic acid, *Planta*, 99, 347, 1971.
6. **Dale, J. W. and Campbell, F.,** Response of tomato plant to stress temperature: increase in abscisic acid concentration, *Plant Physiol.*, 67, 26, 1981.
7. **Horton, D. and Sawyer, R. L.,** The potato as a world food crop, in *Potato Physiology*, Li, P. H., Ed., Academic Press, New York, 1985, 1.
8. **James, R. J. and Mansfield, T. A.,** Suppression of stomatal opening in leaves treated with abscisic acid, *J. Exp. Bot.*, 21, 714, 1970.
9. **Lee, E. H., Byun, J. K., and Steffens, G.L.,** Increased tolerance of plants to SO_2, chilling, and heat stress by a new GA biosynthesis inhibitor, paclobutrazol, *Plant Physiol. Suppl.*, 77, 135, 1985
10. **Levitt, J.,** *Responses of Plants to Environmental Stresses*, Vol. 1, Academic Press, New York, 1980.
11. **Lyons, J. M., Graham, D., and Raison, J. K. Eds.,** *Low Temperature Stress in Crop Plants*, Academic Press, New York, 1979.
12. **Moore, R. and Smith, J. D.,** Growth, graviresponsiveness and abscisic-acid content of *Zea mays* seedlings treated with fluridone, *Planta*, 162, 342, 1984.
13. **Pendleton, J. W. and Egli, D.B.,** Potenital yield of corn as affected by planting date, *Agron. J.*, 61, 70, 1969.
14. **Rao, S. R. and Harger, T. R.,** Mefluidide-bentazon interactions on soybeans and red rice, *Weed Sci.*, 29, 208, 1981.
15. **Rikin, A. and Richmond, A. E.,** Amelioration of chilling injury in cucumber seedlings by abscisic acid, *Physiol. Plant*, 38, 95, 1976.
16. **Rikin, A., Atsman, D., and Gitler, C.,** Chilling injury in cotton (Gossypium hirsutum: preservation by abscisic acid, *Plant Cell Physiol.*, 20, 1537, 1979.
17. **St. John, J. B. and Hilton, J. L.,** Lipid metabolism as a site of herbicide action, *Weed Sci.*, 21, 477, 1973.
18. **Tseng, M. J. and Li, P. H.,** Mefluidide protection of severely chilled crop plants, *Plant Physiol.*, 75, 249, 1984.
19. **Tseng, M. J., Zhang, C. L., and Li, P. H.,** Quantitative measurements of mefluidide protection of chilled corn plants, *J. Am. Soc. Hort. Sci.*, 111, 409, 1986.
20. **Tseng, M. J. and Li, P. H.,** Nucleic acid and protein synthesis during induction of cold hardiness, in *Plant Cold Hardiness*, Li, P. H., Ed., Alan Liss, New York, 1987.
21. **Wright, S. T. C.,** The effect of chilling on ethylene production, membrane permeability and water loss of leaves of *Phaseolus vulgaris*, *Planta*, 120, 63, 1974.
22. **Zhang, C. L., Li, P. H., and Brenner, M. L.,** Relationship between mefluidide treatment and abscisic acid metabolism in chilled corn leaves, *Plant Physiol.*, 81, 699, 1986.
23. **Zhang, L. X., Li, P. H., and Tseng, M. J.,** Amelioration of chilling injury in rice seedlings by mefluidide, *Crop Sci.*, 27, 531, 1987.

Chapter 14

RELATION OF CHILLING STRESS TO ETHYLENE PRODUCTION

Chien Yi Wang

TABLE OF CONTENTS

I. Introduction 178

II. Effect of Chilling Stress on the Pathway of Ethylene Biosynthesis 178

III. Effect of Ethylene Biosynthesis Inhibitors on Chilling-Induced Ethylene Production 179

IV. Timing and Kinetics of Chilling-Induced Ethylene Production 180

V. Use of Endogenous ACC Levels or Ethylene Production as Indicators of Chilling Exposure 182

VI. Effect of Ethylene on the Development of Chilling Injury Symptoms 183

VII. Impairment of Ethylene Biosynthesis by Prolonged Chilling Exposure 184

VIII. Effect of Light on Ethylene and Injury Induced by Chilling 185

IX. Changes of Polyamines During Chilling Stress in Relation to Ethylene Production 185

X. Conclusion 186

References 187

I. INTRODUCTION

Low temperature stress enhances ethylene production of chilling-sensitive plants which do not normally produce significant amounts of ethylene.[1-9] The mechanism of this chilling-induced ethylene production can now be better understood since the biosynthetic pathway of ethylene has become better elucidated. However, because of their diversity, various plant tissues respond differently to chilling stress. Some tissues accelerate their ethylene production during chilling exposure, while others do not show any stimulation until after being transferred to warmer temperatures.

There are many questions concerning chilling-induced ethylene production that need to be answered. Do plants produce ethylene as a defense mechanism against chilling stress or is ethylene production simply a physiological response to stress? Does ethylene play a role in enhancing or reducing the development of symptoms of chilling injury (CI)? How is ethylene biosynthesis stimulated by chilling temperatures? Which steps in the ethylene biosynthetic pathway are affected? What are the kinetics of the stimulation? Can ethylene production be used as an indicator of chilling sensitivity? These and other questions will be discussed in this chapter. While this article does not intend to cover all aspects of the effect of temperature on ethylene production, an effort has been made to summarize various recent findings on chilling-induced ethylene production and to evaluate the implications of these research results. For additional discussion, readers are referred to articles by Chen and Patterson,[10] Field,[11] and Wang and Adams.[12]

II. EFFECT OF CHILLING STRESS ON THE PATHWAY OF ETHYLENE BIOSYNTHESIS

Adams and Yang[13] established that the ethylene biosynthetic pathway in ripening fruits and other higher plants is methionine → S-adenosylmethionine (SAM) → 1-aminocyclopropane-1-carboxylic acid (ACC) → ethylene. One of the first questions concerning chilling stress is whether this same pathway also accounted for the increased production of ethylene from chilled tissues. By feeding L-[3,4-^{14}C]methionine to chilled as well as nonchilled cucumber skin tissues, Wang and Adams[6] found that the radioactive methionine was converted to radioactive ACC in both groups of tissues. The radioactive ACC then yielded labeled C_2H_4 in the chemical degradation assay of Lizada and Yang.[14] The results indicate that methionine is used as the precursor of ethylene in both chilled and nonchilled tissues, and that chilling-induced ethylene is produced through the same pathway as that proposed for ripening fruits.

The second question concerning chilling stress is which steps in the ethylene biosynthetic pathway are affected by chilling and which step is responsible for the stimulation of ethylene production by chilling temperatures? A rapid increase in ACC synthase activity and ACC levels was found in cucumber tissue after the cucumbers were transferred to 25°C following exposure to chilling temperature at 5°C.[12] ACC levels increased fiftyfold within 7 hr of warming after 4 days at 2.5°C. Incorporation of L-[3,4-^{14}C]methionine into tissues of cucumbers resulted in higher levels of labeled ACC and labeled C_2H_4 in chilled tissue than in nonchilled tissue (Table 1), indicating that chilled tissue had a greater capacity for converting methionine to ACC and to ethylene. These data, together with results from studies on inhibitors, described in the next section, suggest that the increased activity of ACC synthase, which leads to the increased formation of ACC, is the step stimulated by chilling stress and is the rate limiting step in chilling-induced ethylene production.

Table 1
[^{14}C]-ACC and [^{14}C]-ETHYLENE PRODUCTION IN SKIN TISSUES OF 2-DAY CHILLED (2.5°C) AND NONCHILLED (13°C) CUCUMBERS AFTER INCUBATION WITH 2 μCi of L-[3,4-^{14}C[METHIONINE FOR 6 HR AT 30°C

Treatment	[^{14}C]-ACC	[^{14}C]-ethylene
	nCi	nCi
Chilled (2.5°C)	4.96	0.285
Nonchilled (13°C)	2.97	0.194

From Wang, C. Y. and Adams, D. O., *Plant Physiol.* 66, 841, 1980. With permission.

Table 2
EFFECT OF VARIOUS INHIBITORS ON ENDOGENOUS ACC LEVEL AND ETHYLENE PRODUCTION IN CHILLED CUCUMBER SKIN TISSUES

			C_2H_4	
Treatment	**Concentration**	**ACC level**	**−ACC**	**+ACC**
	m*M*	**% Control**	**% Control**	
AVG	0.005	39.1	34.9	96.2
	0.05	28.4	25.2	96.7
	0.1	12.6	7.6	91.6
PG	1	120.3	64.9	59.2
SB	1	114.7	73.4	70.1
DNP	0.05	88.6	56.3	49.7
CCCP	0.05	52.4	61.8	43.5
CHI	0.05	46.4	40.7	38.3

Note: Cucumbers were exposed to 2.5°C for 2 days. One g of skin tissue was taken from each of the chilled cucumbers and incubated in 25 mℓ flasks containing 3 mℓ of 10 m*M* Mes buffer (pH 6.0), 2% sucrose, and 100 μg of chloramphenicol. ACC content and ethylene production were analyzed after 2-hr incubation at 30°C. Absolute control values for ACC and ethylene production with and without exogenous ACC (0.1 m*M*) were 3.28 nmol/g, 1.68 and 0.14 nmol/g/hr, respectively.

From Wang, C. Y. and Adams, D. O., *Plant Physiol.*, 66, 841, 1980. With permission.

III. EFFECT OF ETHYLENE BIOSYNTHESIS INHIBITORS ON CHILLING-INDUCED ETHYLENE PRODUCTION

The production of chilling-induced ethylene is affected by several inhibitors (Table 2). Aminoethoxyvinylglycine (AVG) reduces ACC accumulation and inhibits C_2H_4 production in chilled cucumber skin tissues. However, AVG does not inhibit ethylene production when exogenous ACC is added, indicating that AVG inhibits chilling-induced ethylene production by blocking the formation of ACC. This result is consistent with what has been found in other tissues.[15,16]

Free radical scavengers, such as propyl gallate and sodium benzoate, also inhibit chilling-induced ethylene production. However, unlike AVG, propyl gallate and sodium benzoate increase the accumulation of endogenous ACC levels in chilled tissue and inhibit ethylene production, even in the presence of exogenous ACC. This suggests that free radical scavengers do not inhibit the formation of ACC in the chilled tissue, but affect the step between ACC and C_2H_4. Similar inhibition of ethylene production by free radical scavengers was also demonstrated in apple, avocado, tomato, bean, and pea tissues.[17,18]

Uncouplers of oxidative phosphorylation, such as 2,4-dinitrophenol (DNP) and carbonyl cyanide *m*-chlorophenylhydrazone (CCCP), reduce ACC levels and inhibit C_2H_4 production in chilled cucumber skin tissue with or without exogenous ACC. It appears that DNP and CCCP at a concentration of 0.05 m*M* inhibit both the conversion of methionine to ACC and the conversion of ACC to ethylene. Since the conversion of methionine to SAM requires ATP, these uncouplers may decrease the amount of ACC in the tissue by making less ATP available for SAM synthesis. Furthermore, it has also been shown that the synthesis of ethylene from ACC is more vulnerable to DNP inhibition than the synthesis of SAM from methionine.[19] It has been suggested that the conversion of ACC to C_2H_4 may also require ATP; and therefore, is subject to inhibition by uncouplers of oxidative phosphorylation.

In the presence of cycloheximide, both ACC accumulation and ethylene production from exogenous ACC in chilled cucumber skin tissue were inhibited. This indicates that continuous protein synthesis is required for the operation of this pathway during chilling stress. These inhibitor studies demonstrate that the biosynthetic pathway for chilling-induced ethylene production is similar to that in ripening fruits and that the step stimulated by chilling stress is the formation of ACC.

IV. TIMING AND KINETICS OF CHILLING-INDUCED ETHYLENE PRODUCTION

It has been observed that ethylene production in some chilling-sensitive tissues is extremely low at chilling temperatures, but increases rapidly when these tissues are transferred from chilling temperatures to 20 or 25°C. Questions may then be raised as to whether the stimulation of ethylene production occurs during the chilling or warming period, and what the effect is of chilling temperatures on the biosynthesis of ethylene. Since ethylene production is stimulated only after the exposure of tissues to chilling stress, the chilling temperature must have initiated the chain of events, but the later steps leading to the release of ethylene are not completed until after the tissues are transferred to warmer temperatures. It has been postulated that chilling temperatures may induce the formation of a membrane-associated cytoplasmic factor which is responsible for enhanced stress-induced ethylene production after tissue is transferred to 25°C.[2]

Studies with inhibitors of RNA and protein synthesis in chilled tissues offer some insight as to possible mechanisms of ethylene evolution under chilling and warming conditions. Table 3 shows that ACC synthase activity, ACC levels, and ethylene production of the chilled cucumber tissue during warming are inhibited by cycloheximide but not affected by cordycepin or α-amanitin. Since cycloheximide is a protein synthesis inhibitor,[20,21] and cordycepin and α-amanitin are RNA synthesis inhibitors,[22,23] it appears that the steps which are inhibited by the two RNA inhibitors have already been stimulated during the chilling period. The chilling temperature must have induced factors or conditions necessary for the transcriptional processing of RNA synthesis. However, since ACC synthase activity is inhibited by cycloheximide, it is apparent that the final step in the translational process of protein synthesis has not been complete during chilling.

Table 3
EFFECT OF CYCLOHEXIMIDE, CORDYCEPIN AND α-AMANITIN ON ACC SYNTHASE, ACC LEVEL, AND ETHYLENE PRODUCTION IN CUCUMBER SKIN TISSUES AFTER TRANSFER FROM 2.5°C to 30°C

Treatment	ACC synthase (nmol ACC/mg protein-hr)	ACC level (nmol/g)	C_2H_4 (nmol/g-hr × 10^2)
Control	0.32	16.5	25.8
Cycloheximide	0.05	3.8	5.4
Cordycepin	0.34	16.7	25.8
α-amanitin	0.35	17.1	26.6

Note: Cucumbers were chilled at 2.5°C for 4 days. Cycloheximide (0.05 m*M*), cordycepin (0.05 m*M*), and α-amanitin (0.05 m*M*) were vacuum infiltrated into the tissues before transfer of the skin samples to 30°C. Development of ACC synthase, formation of ACC, and production of ethylene were measured after 3 hr incubation of samples with or without inhibitors in 10 m*M* Mes buffer (pH 6.0) and 2% sucrose at 30°C.

From Wang, C. Y. and Adams, D. O., *Plant Physiol.*, 69, 424, 1982. With permission.

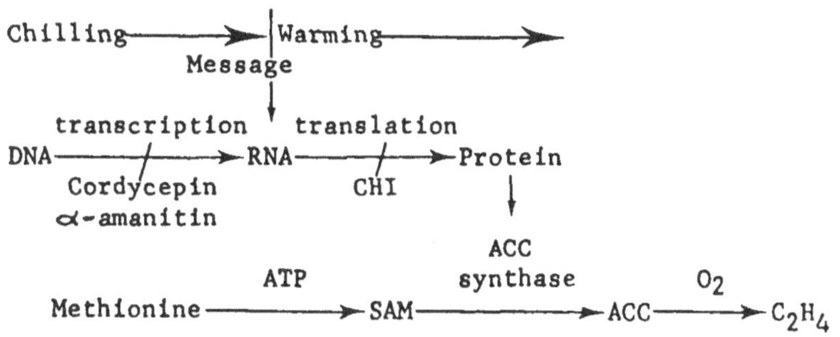

FIGURE 1. Possible mechanism of the effect of chilling stress on the formation of ACC synthase and the production of ethylene.

A simplified scheme (Figure 1) is proposed to illustrate the possible effect of chilling stress on the formation of ACC synthase and the production of ethylene. It is postulated that a message is produced or unmasked during the chilling period. This message probably includes mRNA coding for ACC synthase. However, the translation and the synthesis of new proteins does not occur until after the transfer of tissue to warm temperatures. Although the signal is turned on by chilling stress, ethylene is not produced until after the formation of ACC synthase at warm temperatures.

The above observation that ACC synthesis does not occur during the initial incubation at chilling temperatures but takes place after the transfer of tissues to warmer temperatures has been reported in cucumber fruit[12,24,25] and in the leaf tissue of dwarf beans (*Phaseolus vulgaris* L.).[2,26] Wang and Adams[12] found that ACC synthase activity increased rapidly only upon the transfer of chilled (2.5°C) cucumber tissue to a warmer temperature (25°C). Andersen and Kent[24,25] reported that increases in ACC content and ethylene evolution in cucumber fruit tissue were only evidenced upon warming of the fruit from 2 to 25°C. Field[2,26] treated leaves of dwarf beans with 1.0 m*M* AVG at the end of a chilling period and found that little ethylene was produced even after the transfer of leaf discs to 25°C, indicating that ACC did not accumulate during the initial incubation period at 5°C. He also reported that the addition of exogenous ACC to leaf discs incubated at 5°C increased ethylene production, while similarly incubated leaf discs did not synthesize additional amounts of endogenous ACC until after they were transferred to 25°C.

Accumulation of ACC and an increase in ethylene production at low temperatures, however, have been reported in other tissues. An increase in ACC levels occurred at 0°C in leaves of *Episcia reptans*.[10] ACC levels in the skin and cortex of Honey Dew melons also increased during chilling at 2.5°C, and the capacity of melons to produce ethylene and CO_2 at 20°C was enhanced by chilling temperatures.[27] Likewise, exposure to low temperatures stimulated ethylene synthesis in pears.[28-32] The increased accumulation of endogenous ACC in pears during low temperature storage was thought to be responsible for the increased capability of pears to produce ethylene and ripen.[33-35] Pear fruit develop core breakdown and abnormal ripening after prolonged cold storage.[36,37] These symptoms are similar to those caused by CI.

Further studies are needed to determine the reason why some tissues are able to synthesize ACC at chilling temperatures while others can only accomplish that after being transferred to warmer temperatures. The basic mechanisms underlying these differences in various tissues in their responses to chilling temperatures warrant further investigation.

V. USE OF ENDOGENOUS ACC LEVELS OR ETHYLENE PRODUCTION AS INDICATORS OF CHILLING EXPOSURE

The augmentation of ACC levels and ethylene production is a common response of chilling-sensitive tissues to low temperature stress. Therefore, changes in these compounds have been suggested by several researchers to be used as indicators of chilling sensitivity and to gauge the degree of chilling exposure.

In a study on plants with different chilling sensitivities, Chen and Patterson[10] found that a response to chilling, expressed as an increase in ethylene production and ACC accumulation, was detected within an hour in leaves of a very chilling-sensitive tropical herb *E. reptans*. They also showed that the duration of chilling required for the burst of ethylene production agreed with the ranking of the chilling susceptibility in six species of *Passiflora*. In cucumber skin tissue, an increase in ACC levels was detected at 20°C after 24 hr at 2.5°C, or 36 hr at 5 or 7.5°C (Figure 2).[38] Since the nonchilled tissue did not contain any significant amount of ACC, and since ACC levels increased with the increasing length of chilling exposure, the endogenous ACC level was proposed as an index for chilling exposure in cucumbers. In sweet peppers, the length of chilling treatment required to induce increases of ACC and ethylene production was positively correlated with chilling sensitivity of the pods harvested in different seasons. Sweet peppers harvested in the summer were more chilling sensitive, and ACC and ethylene production increased after 32 hr of exposure to 3°C; but those harvested in the fall were less sensitive to chilling, and ACC and ethylene production did not show any increase until after 60 hr of exposure to 3°C.[39] ACC levels in the rind from the ground spot of Honey Dew melons were consistently higher than those from the top of melons after chilling, and the susceptibility to CI was also higher on the bottom than on the top of the melons.[40] Incipient CI, such as pitting or discoloration, was confined to the skin of cucumbers or the rind of Honey Dew melons. ACC accumulation after chilling stress was also higher in the skin or rind than in the cortex of these fruits;[6,27] and thus, paralleled the sensitivity of these tissues to the development of symptoms of CI. Papaya fruit stored at chilling temperature (5°C) for more than 4 days also produced more ethylene upon transfer to warmer temperatures than did fruit stored at a nonchilling temperature (10°C).[41] Differences in ethylene production between chilled fruit and nonchilled fruit increased with the length of storage. Chan et al.[41] found that an enhanced capacity to produce ethylene existed before the onset of both electrolyte leakage and measurable skin color differences in injured papaya, and concluded that the determination of ethylene ac-

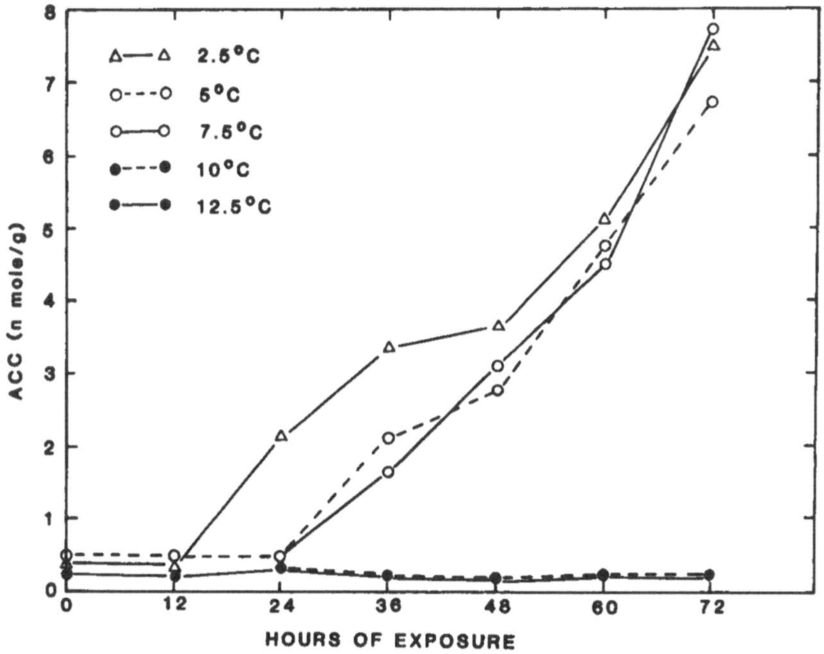

FIGURE 2. Stimulation of ACC levels by various chilling temperatures in cucumber skin tissue. ACC levels were analyzed 7 hr after transfer of tissue to 20°C.

cumulation may provide an improved objective method to determine the onset of CI. Both ACC and ethylene can easily be measured with established techniques. Therefore, using the changes in ACC or ethylene concentration as a method of measuring chilling stress can be very efficient for screening different genotypes for their chilling sensitivities.

The amount of ACC accumulated as 25°C in cucumber skin tissue following chilling showed a continuous increase as the length of chilling exposure increased (Figure 2). However, after 48 hr of chilling treatments, ACC levels in tissues exposed to different chilling temperatures became indistinguishable. Fruit exposed to lower chilling temperatures did not necessarily accumulate higher levels of ACC. In Honey Dew melons, none of the external characteristics that were closely associated with the development or expression of CI, i.e., the degree of ripeness of the melons, the reddish-tan surface discoloration, and the presence of solar yellowing was invariably reflected in the magnitude of ACC accumulation in the melons.[27] Thus, additional factors other than chilling must be involved in determining the course of ACC metabolism in the tissues. These observations suggest that caution should be taken when using this method to assess CI, because the absolute value of ACC may not always reflect the severity of the injury.

VI. EFFECT OF ETHYLENE ON THE DEVELOPMENT OF CHILLING INJURY SYMPTOMS

Does ethylene play a role in enhancing or reducing the development of CI symptoms? If ethylene is directly involved in causing CI, then inhibition of its production or action should lessen the severity of CI. Treatment of chilling-sensitive cucumber seedlings with aminooxyacetic acid, an inhibitor of ethylene biosynthesis, inhibited chilling-induced ethylene production but did not reduce CI of seedlings exposed to 5°C.[42] Application with inhibitors of ethylene action, such as silver thiosulfate[43] or norbornadiene,[44] also did not affect the development of CI symptoms in chilling-sensitive bean plants.[45] These data indicate that ethylene is not a primary cause of CI nor an essential part of the developmental process for the CI symptoms.

If the plants produce ethylene as a means of defense against CI, then augmentation of ethylene should reduce the extent of the injury. Treatment of cucumber seedlings with ACC or ethylene before, during or after chilling exposure did not alleviate CI.[46] Gassing of mature green tomatoes with ethylene before or after storage at a chilling temperature also did not affect the development of CI symptoms.[47,48] Ethylene enhanced softening in chilling-sensitive cultivars of tomatoes only at 20°C.[49] Furthermore, exposure of avocados to ethylene at temperatures below 12°C greatly increased the severity of CI.[50-52]

On the other hand, reduction of CI by ethylene has been reported. Ethylene gassing of Honey Dew melons at 20°C before storage at 2.5°C substantially reduced development of CI.[53] Ethylene treatment was also able to permit a higher rate of respiration and ripening to occur in the melons and led to subsequent normal ripening at 20°C.[54] Degreening of grapefruit with ethylene prior to cold storage has also been found to reduce CI.[55] Exposure of sweet potatoes to ethylene during the post-chilling period significantly reduced the severity of CI.[56] It appears that ethylene may alter the physiology of some crops and indirectly change the sensitivity of these crops to CI. Whether the effect is beneficial or detrimental depends on the type of tissue involved.

VII. IMPAIRMENT OF ETHYLENE BIOSYNTHESIS BY PROLONGED CHILLING EXPOSURE

Chilling-induced ethylene production was found to be reduced after prolonged duration of chilling exposure in cucumber fruit,[12,57] winged bean and yard long bean,[58] and leaves of *E. reptans*.[10] Apparently, the synthetic pathway of ethylene is impaired by severe chilling treatment. Wang and Adams[12] showed that ACC content continued to accumulate in cucumber skin tissue even after the decline of ethylene production following prolonged chilling exposure. Similarly, Andersen[57] found that the conversion of ACC to ethylene in cucumbers decreased as the chilling period increased. Lizada et al.[58] reported that ACC accumulated while ethylene production declined in pods of winged bean and yard long bean as CI symptoms became more pronounced with time of storage at 5°C. Chen and Patterson[10] also reported that while ethylene production in leaves of *E. reptans* was reduced after serious CI, the production of ACC was not reduced. These data indicate that the system which converts ACC to ethylene is impaired by severe chilling.

Table 4 shows that the nonchilled cucumber tissue continued to respond to the application of exogenous ACC even after 10 days of storage at 13°C, but stimulation of ethylene production by exogenous ACC in chilled tissue decreased after 4 days of exposure to 2.5°C. Ethylene production from nonchilled samples remained unchanged throughout the 10-day period both in the presence and absence of exogenous ACC, whereas the capacity of chilled tissue to convert ACC to ethylene decreased with the increasing length of chilling exposure after 4 days. Failure of chilled tissue to respond to exogenous ACC after prolonged chilling suggests that the reduced ethylene production is not because of the absence of ACC in the tissue but due to the loss of capacity to convert ACC to ethylene.

It has been suggested that the conversion of ACC to ethylene requires membrane integrity and is very sensitive to membrane disruption.[59,60] Since an alteration in membrane composition has been known as one of the primary changes induced by chilling,[61,62] it is not surprising that the conversion of ACC to ethylene is the first step to be impaired by chilling temperatures.

Table 4
ETHYLENE PRODUCTION OF CUCUMBER SKIN TISSUE INCUBATED WITH OR WITHOUT EXOGENOUS ACC (0.1 mM) AFTER VARYING DEGREES OF EXPOSURE TO CHILLING (2.5°C) AND NONCHILLING (13°C) TEMPERATURES

	C_2H_4 production (nmol/g-hr × 10^2) days of exposure				
Treatment	**2**	**4**	**6**	**8**	**10**
2.5°C	20.4	27.9	10.0	3.3	1.2
2.5°C + ACC	45.0	46.6	22.1	5.0	1.6
13°C	8.3	9.5	7.5	7.5	8.3
13°C + ACC	47.9	45.8	47.9	42.9	50.8

Note: Samples of 1-g skin tissue were incubated at 30°C in 25-mℓ flasks containing 3 mℓ mM Mes buffer (pH 6.0) and 2% sucrose. Ethylene production was determined after 3 hr of incubation.

From Wang, C. Y. and Adams, D. O., *Plant Physiol.*, 69, 424, 1982. With permission.

VIII. EFFECT OF LIGHT ON ETHYLENE AND INJURY INDUCED BY CHILLING

Light has been found to inhibit the conversion of ACC to ethylene in green leaves of tobacco, oats, and rice.[63,64] It was suggested that the inhibition might be related to the photosynthetic system[63] and mediated through carbon dioxide and the activity of the enzyme converting ACC to ethylene.[64]

In studying the chilling-induced photooxidative leaf damage, Horvath and van Hasselt[65] showed that the degradation of phosphatidyl choline, ethanolamine, and phosphatidyl glycerol occurred during chilling under light. Photooxidative degradation of chlorophyll and polyunsaturated fatty acid also took place during chilling under light. Light has also been reported to accentuate CI through the inactivation of thylakoids and the rapid loss of proton uptake,[66] along with the reduction of photosynthetic capacity.[67]

In a study with low and high photosynthetic photon flux (PPF), chilled cucumber seedlings were found to produce comparable levels of ACC under both PPF levels, though CI symptoms were more severe under high PPF than under low PPF.[68] This result further supports the concept that the absolute values of ACC cannot be used to indicate the degree of severity of CI.

IX. CHANGES OF POLYAMINES DURING CHILLING STRESS IN RELATION TO ETHYLENE PRODUCTION

Since ethylene biosynthesis and polyamine biosynthesis use SAM as a common precursor[69,70] and produce 5′-methylthioadenosine as a common by-product,[60,71] it seems logical to suspect that each pathway, when stimulated or suppressed, tends to interfere with the other. There are examples that this might be the case, such as the promotion of spermine biosynthesis by application of an ethylene inhibitor, aminooxyacetic acid, during the senescence of carnation flowers.[72] However, in the case of chilling stress, production of ACC and ethylene and formation of polyamines were simultaneously stimulated in the chilling-sensitive cu-

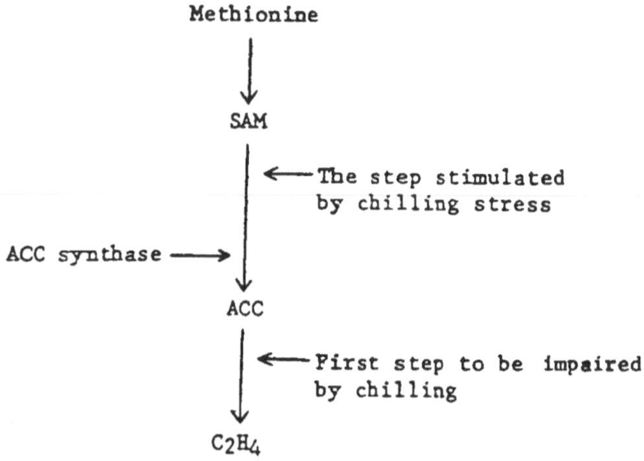

FIGURE 3. The steps affected by chilling stress in the ethylene biosynthetic pathway.

cumber seedlings.[42] Increased accumulation of polyamines after chilling has also been reported in fruits of lemon, grapefruit, and pepper,[73] and leaves of bean.[74] Moreover, although chilling-induced ethylene production in cucumber seedlings was inhibited by aminooxyacetic acid, changes in polyamine levels during chilling were not affected.[42] These results suggest that these two biosynthetic pathways operate independently of each other.

Despite the similarity in their biosynthetic pathways, ethylene and polyamines have opposite functions in plant tissues. Ethylene tends to enhance senescence[60] whereas polyamines tend to stabilize membranes and inhibit senescence.[75,76] Since chilling temperatures enhance the synthesis of both ethylene and polyamines, it is apparent that their common precursor, SAM, does not become a limiting factor for their biosynthesis under chilling stress. Whether the increased polyamine levels in the chilled tissues is correlated with chill tolerance[74] or is a cause of CI[73] warrants further study.

X. CONCLUSION

During the past few years, we have gained additional insight as to the relationship between chilling stress and ethylene production.[77] The biosynthetic pathway for chilling-induced ethylene production was proved to be the same as that in ripening fruits. The step in the ethylene biosynthetic pathway enhanced by chilling stress was found to be the synthesis of ACC, and the first step impaired by severe chilling was the conversion of ACC to ethylene (Figure 3),

In some tissues, stimulation of ACC synthesis and ethylene production did not occur until after the chilled tissues were transferred to a warmer nonchilling temperature. In these tissues, a message was apparently produced or unmasked during the chilling, but the translation of message and the formation of ACC synthase were not completed until after the warming. However, in other tissues, ACC formation and ethylene production is stimulated during chilling. Further studies are needed to determine differences in response of various tissues to chilling stress. Studies on the origin of ethylene and the molecular processes involved in the induction of ethylene production will be helpful in the understanding of how chilling temperatures turn on ethylene biosynthesis and why different tissues respond differently.

There seems to be a commonality between the effect of chilling stress and alteration in ethylene biosynthesis. Changes in membrane integrity appear to be associated with both the CI in general and with the step where impairment occurs in the ethylene biosynthetic pathway.

The increase in ACC levels and ethylene production can be used as indicators of chilling exposure, but their absolute values do not reflect the degree of CI severity. The stimulation of ethylene production appears to be simply a response of tissues to chilling stress. Ethylene is not a cause of CI or a defense mechanism produced by plant tissues, because the inhibition of chilling-induced ethylene production by ethylene biosynthesis inhibitors does not affect the development of CI symptoms. Treatment of plant tissues with ethylene can either reduce or increase CI, depending on the type of tissue involved. The role of ethylene in the development of CI requires further study.

Chilling stress increases both polyamine and ethylene biosyntheses. Inhibition of chilling-induced ethylene production does not affect changes of polyamines during chilling stress. It appears that a plentiful amount of SAM is constantly available in the tissue to serve as substrates for both polyamine and ethylene biosyntheses and the two biosyntheses are not competitive with each other under chilling stress.

ACC synthase activity, ACC formation, and ethylene production in chilling-sensitive tissues are readily stimulated by low temperature stress. The kinetics of the induction is different in different tissues. Chilling stress, thus, offers an ideal system for the studies of changes which lead to an increase in ethylene biosynthesis.

REFERENCES

1. **Cooper, W. C., Rasmussen, G. K., and Waldon, E. S.,** Ethylene evolution stimulated by chilling citrus and Persea sp., *Plant Physiol.*, 44, 1194, 1969.
2. **Field, R. J.,** The effect of low temperature on ethylene production by leaf tissue of *Phaseolus vulgaris* L., *Ann. Bot.*, 47, 215, 1981.
3. **Ichii, T. and Hamada, K.,** Studies of "rind yellow spot", a physiological disorder of Naruto (*Citrus medioglobosa* Tanaka): low temperature and ethylene evolution from injured fruits, *J. Jpn. Soc. Hortic. Sci.*, 47, 1, 1978.
4. **Apeland, J.,** Factors affecting nonparasitic disorders of the harvested product of cucumber, *Acta Hort.*, 4, 102, 1966.
5. **Vine, H. M., Grierson, W., and Edwards, G. J.,** Respiration, internal atmosphere, and ethylene evolution of citrus fruits, *Proc. Am. Soc. Hortic. Sci.*, 92, 227, 1968.
6. **Wang, C. Y.,** Ethylene production by chilled cucumbers (*Cucumis sativus* L.), *Plant Physiol.*, 66, 841, 1980.
7. **Wright, M.,** The effect of chilling on ethylene production, membrane permeability and water loss of leaves of *Phaseolus vulgaris*, *Planta*, 120, 63, 1974.
8. **Olorunda, A. O. and Looney, N. E.,** Response of squash to ethylene and chilling, *Ann. Appl. Biol.*, 87, 465, 1977.
9. **Eaks, I. L.,** Effect of chilling on respiration and volatiles of California lemon fruit, *J. Am. Soc. Hortic. Sci.*, 105, 865, 1980.
10. **Chen, Y. Z. and Patterson, B. D.,** Ethylene and 1-aminocyclopropane-1-carboxylic acid as indicators of chilling sensitivity in various plant species, *Aust. J. Plant Physiol.*, 12, 377, 1985.
11. **Field, R. J.,** The effect of temperature on ethylene production by plant tissues, in *Ethylene in Plant Development*, Roberts, J.A. and Tucker, G.A., Eds., Butterworths, London, 1985, 47.
12. **Wang, C. Y. and Adams, D. O.,** Chilling-induced ethylene production in cucumbers (*Cucumis sativus* L.), *Plant Physiol.*, 69, 424, 1982.
13. **Adams, D. O. and Yang, S. F.,** Ethylene biosynthesis: identification of 1-aminocyclopropane-1-carboxylic acid as an intermediate in the conversion of methionine to ethylene, *Proc. Natl. Acad. Sci. U.S.A.*, 76, 170, 1979.
14. **Lizada, M. C. C. and Yang, S. F.,** A simple and sensitive assay for 1-aminocyclopropane-1-carboxylic acid, *Anal. Biochem.*, 100, 140, 1979.
15. **Boller, T., Herner, R. C., and Kende, H.,** Assay for and enzymatic formation of an ethylene precursor, 1-aminocyclopropane-1-carboxylic acid, *Planta*, 145, 293, 1979.
16. **Yu, Y. B., Adams, D. O., and Yang, S. F.,** 1-Aminocyclopropane-carboxylate synthase, a key enzyme in ethylene biosynthesis, *Arch. Biochem. Biophys.*, 198, 280, 1979.

17. **Apelbaum, A., Wang, S. Y., Burgoon, A. C., Baker, J. E., and Lieberman, M.,** Inhibition of the conversion of 1-aminocyclopropane-1-carboxylic acid to ethylene by structural analogs, inhibitors of electron transfer, uncouplers of oxidative phosphorylation, and free radical scavengers, *Plant Physiol.*, 67, 74, 1981.
18. **Baker, J. E., Lieberman, M., and Anderson, J. D.,** Inhibition of ethylene production in fruit slices by a rhizobitoxine analog and free radical scavengers, *Plant Physiol.*, 61, 886, 1978.
19. **Yu, Y. B., Adams, D. O., and Yang, S. F.,** Inhibition of ethylene production by 2,4-dinitrophenol and high temperature, *Plant Physiol.*, 66, 286, 1980.
20. **Ellis, R. J., and MacDonald, I. R.,** Specificity of cycloheximide in higher plant systems, *Plant Physiol.*, 46, 227, 1970.
21. **Pestka, S.,** Inhibitors of protein synthesis, in *Molecular Mechanisms of Protein Biosynthesis*, Weissbach, H. and Pestka, S., Eds., Academic Press, New York, 1977, 467.
22. **Strain, G. D., Mullinix, L., and Bogorad, L.,** RNA polymerases of maize: nuclear RNA polymerases, *Proc. Natl. Acad. Sci. U.S.A.*, 68, 2647, 1971.
23. **Takegami, T. and Yoshida, K.,** Remarkable retardation of senescence of tobacco leaf disc by cordycepin, an inhibitor of RNA polyadenylation, *Plant Cell Physiol.*, 16, 1163, 1975.
24. **Andersen, C. R. and Kent, M. W.,** Storage of cucumber fruit at chilling temperatures, *HortScience*, 17, 527, 1982.
25. **Andersen, C. R. and Kent, M. W.,** Chilling injury of cucumber fruit, *Plant Physiol.*, 72 (Suppl.), 95, 1983.
26. **Field, R. J.,** The role of 1-aminocyclopropane-1-carboxylic acid in the control of low temperature induced ethylene production in leaf tissue of *Phaseolus vulgaris* L., *Ann. Bot.*, 54, 61, 1984.
27. **Lipton, W. J. and Wang, C. Y.,** Chilling exposures and ethylene treatment change the level of 1-aminocyclopropane-1-carboxylic acid in "Honey Dew" melons, *J. Am. Soc. Hortic. Sci.*, 112, 109, 1987.
28. **Ulrich, R.,** Temperature and maturation: pears require preliminary cold treatment, *Recent Adv. Bot.*, 1961, 1172, 1961.
29. **Hansen, E.,** Postharvest physiology of fruits, *Annu. Rev. Plant Physiol.*, 17, 459, 1966.
30. **Wang, C. Y., Mellenthin, W. M., and Hansen, E.,** Effect of temperature on development of premature ripening in "Bartlett" pears, *J. Am. Soc. Hortic. Sci.*, 96, 122, 1971.
31. **Looney, N. E.,** Interaction of harvest maturity, cold storage and two growth regulators on ripening of Bartlett pears, *J. Am. Soc. Hortic. Sci.*, 97, 81, 1972,
32. **Sfakiotakis, E. M. and Dilley, D. R.,** Induction of ethylene production in "Bosc" pears by postharvest cold stress, *HortScience*, 9, 336, 1974.
33. **Wang, C. Y., Sams, C. E., and Gross, K. C.,** Ethylene, ACC, soluble polyuronide, and cell wall noncellulosic neutral sugar content in "Eldorado" pears during cold storage and ripening, *J. Am. Soc. Hortic. Sci.*, 110, 687, 1985.
34. **Blankenship, S. M. and Richardson, D. G.,** Development of ethylene biosynthesis and ethylene-induced ripening in "d'Anjou" pears during the cold requirement for ripening, *J. Am. Soc. Hortic. Sci.*, 110, 520, 1985.
35. **Blankenship, S. M. and Richardson, D. G.,** ACC and ethylene levels in "d'Anjou" pears in air and controlled-atmosphere storage, *HortScience*, 21, 1020, 1986.
36. **Claypool, L. L.,** Further studies on controlled atmosphere storage of "Bartlett" pears, *J. Am. Soc. Hortic. Sci.*, 98, 289, 1973.
37. **Wang, C. Y.,** Pear fruit maturity, harvesting, storage, and ripening, in *The Pear*, van der Zwet, T. and Childers, N. F., Eds., Horticulture Publications, Gainesville, Fla., 1982, 431.
38. **Wang, C. Y.,** Use of 1-aminocyclopropane-1-carboxylic acid level as an index for chilling exposure at various temperatures, *Plant Physiol.*, 72 (suppl.), 43, 1983.
39. **Yao, K., Yu, L., and Zhou, S.,** A study of storage temperature and chilling injury on sweet pepper, *Acta Hortic. Sinica*, 13, 119, 1986.
40. **Lipton, W. J., Peterson, S. J., and Wang, C. Y.,** Solar radiation influences solar yellowing, chilling injury and ACC accumulation in "Honey Dew" melons, *J. Am. Soc. Hortic. Sci.*, 112, 503, 1987.
41. **Chan, H.T., Sanxter, S., and Couey, H. M.,** Electrolyte leakage and ethylene production induced by chilling injury of papayas, *HortScience*, 20, 1070, 1985.
42. **Wang, C. Y.,** Changes of polyamines and ethylene in cucumber seedlings in response to chilling stress, *Physiol. Plant.*, 69, 253, 1987, (In press).
43. **Veen, H.,** Silver thiosulphate: an experimental tool in plant science, *Sci. Hortic.*, 20, 211, 1983.
44. **Sisler, E. C. and Yang, S. F.,** Anti-ethylene effects of *cis*-2-butene and cyclic olefins, *Phytochemistry*, 23, 2765, 1984.
45. **Tong, C. B.,** Induction and Characterisitics of Stress Ethylene Biosynthesis, Ph.D. thesis, University of California, Davis, 1986.
46. **Wang, C. Y.,** unpublished data, 1987.
47. **Kader, A. A. and Morris, L. L.,** Amelioration of chilling injury symptoms on tomato fruits, *HortScience*, 10, 324, 1975.

48. **Ogura, N., Hayashi, R., Ogishima, T., Abe, Y., Nakagawa, H., and Takehana, H.**, Ethylene production by tomato fruits at various temperatures and effect of ethylene on the fruits, *Nippon Nogeikagaku Kaishi*, 50, 519, 1976.
49. **Manzano-Mendez, J., Hicks, J., and Masters, J. F.**, Influence of storage temperature and ethylene on firmness, acids, and sugars of chilling-sensitive and chilling-tolerant tomato, *J. Am. Soc. Hortic. Sci.*, 109, 273, 1984.
50. **Young, R. E. and Lee, S. K.**, The effect of ethylene on chilling and high temperature injury of avocado fruit, *HortScience*, 14, 421, 1979.
51. **Lee, S. K. and Young, R. E.**, Temperature sensitivity of avocado fruit in relation to C_2H_4 treatment, *J. Am. Soc. Hortic. Sci.*, 109, 689, 1984.
52. **Chaplin, G. R., Wills, R. B. H., and Graham, D.**, Induction of chilling injury in stored avocados with exogenous ethylene, *HortScience*, 18, 952, 1983.
53. **Lipton, W. J. and Aharoni, Y.**, Chilling injury and ripening of "Honey Dew" muskmelons stored at 2.5° or 5° after ethylene treatment at 20°C, *J. Am. Soc. Hortic. Sci.*, 104, 327, 1979.
54. **Lipton, W. J., Aharoni, Y., and Elliston, E.**, Rates of CO_2 and ethylene production and of ripening of "Honey Dew" muskmelons at a chilling temperature after pretreatment with ethylene, *J. Am. Soc. Hortic. Sci.*, 104, 846, 1979.
55. **Grierson, W.**, Chilling injury in tropical and subtropical fruit. V. Effect of harvest date, degreening, delayed storage and peel color on chilling injury of grapefruit, *Proc. Trop. Reg. Am. Soc. Hortic. Sci.*, 18, 66, 1974.
56. **Buescher, R. W.**, Hard core in sweet potato roots as influenced by cultivar, curing, and ethylene, *HortScience*, 12, 326, 1977.
57. **Andersen, C. R.**, Rapid synthesis of ACC and ethylene by chilling-stressed cucumber fruit, *HortScience*, 21, 858, 1986.
58. **Lizada, M. C. C., Sjaifullah, Agillon, A. B., and Valdez, E. R.**, 1-aminocyclopropane-1-carboxylic acid in tropical fruits and vegtables, *Acta Hortic.*, 138, 329, 1983.
59. **Lieberman. M.**, Biosynthesis and action of ethylene, *Annu. Rev. Plant Physiol.*, 30, 533, 1979.
60. **Yang, S. F. and Hoffman, N. E.**, Ethylene biosynthesis and its regulation in higher plants, *Annu. Rev. Plant Physiol.*, 35, 155, 1984.
61. **Lyons, J. M.**, Chilling injury in plants, *Annu. Rev. Plant Physiol.*, 24, 445, 1973.
62. **Wang, C. Y.**, Physiological and biochemical responses of plants to chilling stress, *HortScience*, 17, 173, 1982.
63. **Gepstein, S. and Thimann, K. V.**, The effect of light on the production of ethylene from 1-aminocyclopropane-1-carboxylic acid by leaves, *Planta*, 149, 196, 1980.
64. **Kao, C. H. and Yang, S. F.**, Light inhibition of the conversion of 1-aminocyclopropane-1-carboxylic acid to ethylene in leaves is mediated through carbon dioxide, *Planta*, 155, 261, 1982.
65. **Horvath, I. and van Hasselt, P. R.**, Inhibition of chilling-induced photooxidative damage to leaves of *Cucumis sativus* L. by treatment with amino alcohols, *Planta*, 164, 83, 1985.
66. **Garber, M. P.**, Effect of light and chilling temperatures on chilling-sensitive and chilling-resistant plants, *Plant Physiol.*, 59, 981, 1977.
67. **Taylor, A. O. and Rowley, J. A.**, Plants under climatic stress. I. Low temperature, high light effects on photosynthesis, *Plant Physiol.*, 47, 713, 1971.
68. **Wang, C. Y.**, Effect of temperature and light on ACC and MACC in chilled cucumber seedlings, *Sci. Hortic.*, 30, 47, 1986.
69. **Adams, D. O. and Yang, S. F.**, Methionine metabolism in apple tissue, *Plant Physiol.*, 60, 892, 1977.
70. **Cohen, S. S.**, *Introduction to the Polyamines*, Prentice-Hall, Englewood Cliffs. N.J., 1971, 1.
71. **Wang, S. Y., Adams, D. O., and Lieberman, M.**, Recycling of 5′-methylthioadenosine-ribose carbon atoms into methionine in tomato tissue in relation to ethylene production, *Plant Physiol.*, 70, 117, 1982.
72. **Roberts, D. R., Walker, M. A., Thompson, J. E., and Dumbroff, E. B.**, The effects of inhibitors of polyamine and ethylene biosynthesis on senescence, ethylene production and polyamine levels in cut carnation, *Plant Cell Physiol.*, 25, 315, 1983.
73. **McDonald, R. E. and Kushad, M. M.**, Accumulation of putrescine during chilling injury of fruits, *Plant Physiol.*, 82, 324, 1986.
74. **Guye, M. G., Vigh, L., and Wilson, J. M.**, Polyamine titre in relation to chill-sensitivity in *Phaseolus* sp., *J. Exp. Bot.*, 37, 1036, 1986.
75. **Galston, A. W.**, Polyamines as modulators of plant development, *BioScience*, 33, 382, 1983.
76. **Smith, T. A.**, Polyamines, *Annu. Rev. Plant Physiol.*, 36, 117, 1985.
77. **Lipton, W. J.**, Postharvest horticulture of vegetables and melons - retrospect and prospect, *Proc. 21st Intl. Hortic. Congr.*, 2, 591, 1982.

Index

INDEX

A

ABA
  the adaptive response to and, 4
  ATP accumulation as growth regulator and, 33
  effects on acclimation and, 171—172
  effects of ethanol and, 11, 12
  effects of stress and, 30—31
  effects of variable photoperiods and, 5
  extracellular proteins and, 20
  optimum concentration of and, 9, 10
  potato and, 5
ABA, endogenous, 5, 172—174
ABA, exogenous,
  effects on cold hardening and, 173
  prevention of chilling injury and, 5—6
  responses to and, 8—9
  variable levels in cold vs. warm crops and, 5—6
ABA/GA ratio, 30
ABA induced proteins, 20
ABA removal, 13
Abiotic/biotic winter damage conditions, 94
Abiotic elicitors, 71—72
Abscisic acid, see ABA
Abscission of leaves, 83
ACC
  effects on ethylene production and, 178—179, 184
  glycoprotein synthesis and, 82
  as indicator of chilling exposure, 182—183
  as metabolized from lipoperoxide, 36
Acclimation, see also Cold hardening
  adaptations of cell constituents and, 28
  ethylene production and, 84—86
  glucose metabolism and, 33—34
  growth regulators and, 4—7
  metabolism modification and, 29—35
  protein synthesis and, 16
  resistance to cellulase and, 74—75
  suberization and, 70—71
  temperature changes and, 4
ACC synthase, 84, 178—182
Acetaldehyde, 112
Acid phosphatase changes, 34—35
Activation energy, 33
Active transport system, see also ATPase activity, 124—125
Adaptation
  ABA and, 4
  of Arctic plants, 92
  injury resistance of winter cereals and, 108—119
  response of wheat and, 117
Adenine nucleotides, 33—34
Adenylate energy charge, see AEC
Adenylate reduction charges, 37
Adventitious root formation, 83
AEC (Adenylate energy charge), 118
Air pollutants, 82
Air and soil temperature variations, 4—5
Alanine
  cellular osmo regulation and, 31—32
  effects of flooding and, 112
  metabolism modification and, 30, 37
  NADPH and, 31
Alcohol dehydrogenase, 117
Alfalfa
  ABA applications to prevent chilling injury and, 5, 6, 9
  effects of ethylene on acclimation and, 85
  GA levels and, 6
  ice encasement and, 115
Almond, 72
Alternative electron pathway, 34—35
Amide-linked fatty acids, 44
Amino acid, 20, 31—32
Aminocyclopropane-1-carboxylic acid, 36, see also ACC
Aminoethyoxyvinyl-glycine, see AVG
Amphibic cell membrane proteins, 46—47
Anabolic reduction charges, 33—35, 37
Anaerobic stress, 112, 116—117
Anatomical alterations, of plasma membrane, 54
Anoxia, 112
Anthesis, 93—94
Antimicrobial secondary metabolites, 72
Anti-parallel pattern, 58
Apical growth cessation, 98—104
Apple
  cryopreservation with LN and, 143—144
  effects of ethylene on acclimation and, 85, 180
  lignification and, 74
  phytoalexin induction and, 72
Aqueous two-polymer phase system, 42
Arctic ecotypes, 92—94
Arctic/subarctic ecotypes, 92
Arginine amino acids, 20
Arrhenius plots, 48
Asparagine, 112, 116
Aspartate, asparagine, 30, 31, 37
Assimilatory power, 32—33
Atmosphere modification and frost control, 159
ATP accumulation
  energy metabolism and, 35
  growth regulators/metabolic pathway modification and, 33—34, 37
  unsaturated lipids and phospholipids and, 32
ATPase activity
  biochemical changes in plasma membrane during freezing and, 49
  effects of freeze injury and acclimation and, 124—135
  effects on ion transport system and, 118, 135
  surface properties of plasma membrane and, 127
ATPase binding orientation, 129—131
ATP production, 54, 55

Auxins, 6—7
AVG, 84, 85, 179—180
Avocado, 180, 184

B

B-9, 7
Banana, 72
Barley
  ABA applications and, 9
  cytokinins and, 7
  mesocotyl elongation and, 17
  proline accumulation and, 9
  protein synthesis and, 16
Beans
  ethylene production during chilling injury and, 180, 184
  phytoalexin induction and, 72
  suberin and, 71
Beech, 72
Beet, 8
Biannual plants, 36
Bilayer fluidity, plasma membrane, 124
Bilayer lipids, 124
Biochemical changes, 141
Biochemical changes, plasma membrane, 49, 54
Biosynthesis of polyunsaturated fatty acids, 32
Biosynthetic potential preservation, 142
Biotic elicitors, 71—72
Biphasic physiologic response, 34—35
Birdsfoot treefoil, 6
Blue light, 99, 103, 104
Breeding selection criteria, 82
Bromegrass, 5—6
Budbreak of cloudberry, 96
Bud primordia, 153—156
Bud set, 9, 103

C

Cabbage, 7, 125
Cadmium toxicity, 72—73
Caffeic acid, 71
Calcium, 124—125
Calcium balance, 36
Calmodulin, 124—125
Calvin cycle, 31, 35
Carbohydrate levels, 94—96, 116—117
Carbon assimilation pathways, 31
Carbon dioxide accumulation, 115—118
Carbon skeletons, 35
Carbonyl cyanide *m*-chlorophenylhydrazone, see CCCP
Carnation, 146—147
Carrot
  ABA applications and, 8
  cryopreservation and, 141
  ethylene synthesis and, 73
  suberin and, 71
Catabolic reduction in cold hardening, 33
Catechol and fungal resistance, 72
Ca_2 levels and H±ATPases, 124—125
CCCP, 180
Celery, 83
Cell component adjustments, 28
Cell differentiatian, 8—9
Cell enlargement, 29, 86
Cell membrane sensitivity, 36
Cell permeability under ice stress, 117
Cell protein analysis methods, 17—19
Cell sap, 83, 147
Cell shrinkage, 48
Cell suspension cultures, 9—21
Cellular osmo regulation, 31—32
Cell wall changes
  effects of ABA and, 16
  plasma membrane proteins and, 44
  vascular bundles and, 71
Ceramide monohexoside, see Cerebroside
Cerebroside, 42—44
Chemical structure, 44
Cherry, 85, 162
Chilling injury
  effects of ABA and, 171—172
  ethylene production and, 82, 178—187
  measurement of amount and, 182—183
Chilling injury prevention, 6
Chilling injury protection, 168—176
Chill unit accumulation, 160—161
Chlorophyll, 127—131, 185
Chlorophyll-protein complexes, 56—57
Chloroplast membrane, 33
Chloroplasts
  etioplasts and, 57—58
  metabolites and, 31
  PG and, 59
  photosynthetic electron transport and, 110
  trans 16:1 and, 59, 60
Chromosomal abberations, 141
Cinnamate 4-hydroxylase, 70, 86
Cinnamate (cinnamic acid), 70, 75
Cinnamic acid, see Cinnamate
Citrus plants, 85
Climatic growth regulation, 92—104
Clonal propagated plant species, 140—143
Cloudberry, 95—98, 104
Cold hardening, see also Acclimation; Freezing injury, and specific plants
  ABA/GA levels and, 6, 7, 9—14
  auxins and, 6—7
  dark respiration and, 33—34
  energy changes and, 33—34
  enzymatic level changes and, 33—34
  etiolation (dark induced)/variable photoperiods and, 6
  first stage of and, 28, 29
  growth regulation and, 7
  kinetin and, 7
  mRNA transcripts and, 16
  photosynthesis changes and, 33—34
  phytochrome and, 6
  pyridine nucleotides and, 31, 32
  second stage of and, 29
Cold hardening, ABA induced,

effects of cycloheximide and, 16
effects of ethanol and, 11, 12
effects of incubation temperature and, 9—11
effects of inoculum size and, 9, 10
effects of light variables and, 12
effects of nitrogen nutrients and, 11, 12
effects of protein changes and, 16—21
effects of sucrose and, 13
effects of time variables and, 11
Cold hardiness forecasting, 159—162
Cold stress response, 73—75
Concentration of mefluidide, 170, 171
Condensation of solutes, 48
Constitutional adaptations, 35
Cooling rate of ice formation, 154—155
Coomassie blue, 19
Corn, 8, 168—176
Cortex injuries,of woody plants, 101
Cotton, 5, 173
Coumarin (coumaric acid), 68, 72
Covalent coupling, 68
CO_2, 29—30, 112
Critical temperature forecasting of flower buds, 160
Crop improvement by cryopreservation, 141
Crop rotation of winter cereal, 109
Cross-adaptation, 7—8, 14—16
Cryo-chrome path, 34
Cryo-genic storage of seeds, 140—143
Cryo-microscopy of protoplast osmotic behavior, 109
Cryopreservation
droplet freezing and, 147—148
history of and, 140
methodology of and, 143—149
post-thaw regrowth and toxicity and, 148—149
pregrowth conditioning and, 144
rapid freezing and, 146—147
recrystalization and, 148
slow freezing and, 147
Cryoprotectants
categories of and, 145—146
glucose and, 34
protoplasts and, 146
sucrose and, 13
for use with LN and, 143
Cryoprotection, 33
Cryoprotective chemicals, 147
Cucumber
ABA applications and, 9, 173
ACC levels and, 182—184
cross adaptation to stress tolerance and, 7—8
ethylene production and, 178—179
phytoalexin induction and, 72
Culm development, 93—94
Currant, 143—144
C_4-dicarboxylic acid, 31
Cyanide-resident respiratory path, 35
Cyanide-resistant electron path, 37
Cycloheximide, 16, 180
Cycocel, 7
Cytochrome f/b_6 complex, 55
Cytochrome path inhibition, 34
Cytochrome P_{450} oxygenase, 70
Cytokinins, 7
Cytoplasmic lipid-exchange protein, 110
Cytosol, 36

D

Dark respiration, 29—31, 33—34
DCCD, 127
DCCD-sensitivity, 49
Dehardening, 98, 103
Dehydration, see also Water stress
during cryopreservation, 147
effects on cell membrane and, 36
exocytotic extrusions and, 109
plasma membrane damage and, 48
in the third stage of cold hardening, 29
vapor pressure gradient and, 147
Densitometer tracings, 19
Deoxyribonucleotides, 32
Depolarization, 72—73
Detergent-rich vs. detergent poor phase, 46—47
DGDG, 59, 60
DHP-fluorescent polarization, 47
Dicarboxylic acid, 70
Differential scanning calorimetry, 59
Differential thermal analysis, see DTA
Digalactosyldiacyl-glycerol, see DGDG
Digestion by enzymes, 16
Dihydrophaseic acid, 17, see also DPA
Dihydroxy phenolics, 73
Dimeric LHCII, 57—63
Dimethylsulfoxide, see DMSO
Dinitrophenol, see DNP
Diphenyl-hexatriene, see DHP
Disease resistance
ethylene production and, 82
glycoproteins and, 82
phenolic compounds and, 72
phytoalexins and, 72
plasma membrane proteins and, 44
DMSO treatment, 144—146
DNA, 9
DNP, 180
Dormancy, 16
Dormin, see ABA
Douglas fir, 71
DPA, 5
DPH, 42, 48
Droplet freezing, 147—148
Dry matter production, 29, 92—95
DTA
supercooling ability of tissue water and, 28
of supercooling of flower buds, 154—155
supercooling flower buds and, 160
Dual induction requirement, 93—94

E

Economic tree line, 92
Ecotype variables,101—104, 108
Electrolyte leakage, 172

Electron microscopy, 109
Electron spin resonance spectroscopy, 117
Electron transport path, see also Photosynthetic electron transport path
in ATP synthesis, 54
chloroplasts and, 110
cytochrome path inhibition and, 34
mitochondria and, 34
Electrophoresis of cell protein, 17—18
Electrophoretic mobility, 19
Elongation affected by photoperiod variables, 92—94, 99
Elongation of cloudberry, 96
Elongation growth cessation of woody plants, 104
Embark®, see Mefluidide
Embedded fluorophores, 42, 48
Embryo growth, 11
Endocytotic vesiculation, 109
Endogenous hardiness inhibitor/promotor, 5
Energy changes during cold hardening, 25, 33—34
Energy coupling reactions in ATP synthesis, 54
Environmental control of growth, see Climatic growth regulation
Environmental cues, 4
Enzymatic level changes, 34—35
Enzymatic regulation of phenolic biosynthesis, 68—71
Enzyme activity blockage of ion transport system, 118
ER (endomembrane), 42
ER membrane, 70
Ethanolamine, 185
Ethanol levels
effects of flooding and, 112
effects of ice encasement and, 115—118
growth regulation and, 11, 12
Ethelene glycol, 48
Ethephon, 159
Ethylene
effects on PAL synthesis and, 70
as indicator of chilling exposure, 182—183
lignification and, 70
membrane depolarization and, 72—73
as metabolized from lipoperoxide, 36
phenolic compounds and, 73
Ethylene production
effects of chilling injury and, 178—179, 184—186
effects of inhibitors and, 179—180
effects on respiration and, 184
effects of SAM and, 185
effects of temperature changes and, 180—182
stress resistance and, 82—86
Etiolation, 6
Etioplasts, 57—58
Eutectic point of cryopreservation, 148
Evan's blue measurement of cryopreservation regrowth, 149
Exocytotic extrusions, 109
Exotherm of flower buds, 160
Extracellular ice formation, 147, 148
Extracellular proteins, 20
Extraorgan freezing, 155

F

Far-red light, 99, 103
Fatty acid, 86, see Polyunsaturated fatty acids
Fatty acid catabolism, 37
Fatty acid composition
of lipid/phosopholipid bilayers, 42, 44
of PG, 54, 62—63
Fatty acid metabolism, 32, 37
Ferulic acid
fungal resistance and, 72
in microsomal fraction, 70
phenolic compounds and, 75
First phase of cold hardening-insensitive plants, 36
Flavonoid, 68
Flower bud injury temperatures, 162
Fluorescein diacetate, 149
Fluorescent anisotropy parameters, 48—49
Fluorescent polarization
of embedded fluorophore, 42
rotational relaxation time and, 48—49
to measure fatty acid composition of PG, 59
Fluorogram measurement of protein, 18
Fluridone, 174—175
Forage grasses affected by ice encasement, 114
Free radicals
polymerization and, 70
protection by glutathione and, 32
scavenger effects on ethylene production and, 180
Freeze fracture, 109
Freeze injury, see also Cold hardening
effects on ATPase activity and, 132—135
effects on plasma membrane and, 36, 48, 54, 108, 109
ethylene production and, 82, 84—85
intracellular water loss and, 13
ion efflux and, 125
rehydration and, 110
Freeze tolerance of northern grass species, 94
Frost control by atmosphere modification, 159
Frost tolerance development, 29
Fructose accumulation during ice encasement, 116
Fungal infection, 70, 72

G

GA
the adaptive response to and, 4
alfalfa and, 6
cold hardening and, 5, 6, 12—13, 159
growth regulators and, 17
responses to and, 8—9
variable photoperiods and, 6
GA/ABA ratio in photosynthetic pathways, 30
Gene regulation, 17
Genetically heterozygous seeds, 140
Genetic constituents, 17
Gene transcription, 17
Germination, 30, 71
Germplasm cryopreservation, 143—144
Gibberellic acid, see GA
Glucose-6-phosphate dehydrogenase changes, 33—

35
Glucose, see also Sugar
alternative electron path and, 34
as a cryoprotectant, 34, 145—146
osmotic action and, 34
phenolic accumulation and, 72
Glucose catabolism, 33—34, 37
Glucose metabolism, 33—34
Glutamate, 30, 37
Glutathione
cell function maintenance and, 31—32
cold hardening insensitive plants and, 34—35
Glycerol, 86, 145—146
Glycine, 30—32
Glycolic acid pathway, 35—37, see also Photorespiration
Glycolysis, 33
Glycolytic potential, 117
Glycoproteins, 46—47, 82
Gooseberry, 143—144
Grape, 74, 162
Growth cycle/rhythm, 92—95, 104
Growth regulation
ABA applications and, 8—9
ATP accumulation and, 33
auxins and, 6—7
by climate, 92—104
cell performance in low temperature and, 28
effects of GA on cold hardening and, 4—7, 17
effects of mefluidide and, 168, 170
ethanol and, 11, 12
phenolic compounds and, 71

H

Harberlandt acid, 36
H-bonding in protein modification, 68
Heat, 30
Heat stress, 14—16, 72
Heat sum, 92, 99
Herbicide, 70, 82
Hexadecanoic acid, 59, 60
Honey Dew, 182—184
Hormone action of plasma membrane proteins, 44
Horticultural factors in forecasting, 160
H±ATPases, 124—126
Hydrolysis of polysaccarides, 29
Hydrophobic interactions, 68
Hydrophobicity of cell membrane proteins, 46—47
ω-Hydroxy fatty acids, 70
Hydroxylase:*p*:coumarate CoA ligase, 73
Hydroxylated long chain fatty acids, 44
Hydroxyproline, 82
Hypertrophied lenticel, 83
Hypothetic, wound hormone, 36

I

IAA decarboxylation of caffeic acid, 71
IAA-oxidase activity in ethylene synthesis, 73
Ice encasement of winter cereals, 113—119
Ice nuclei/formation in flower buds, 153—154
Immunization, 72
Incubation conditions of protein synthesis, 17
Inflection, 48—49
Inner rhythm of plants during dehardening, 98
Inoculum size, 9, 10
Insect attack, 70, 82
Inside-out ATPase orientation, 131
Intrinsic and extrinsic components of thylakoid membrane, 55
Ion efflux, 125
Ion interactions, 68
Ion leakage, 134
Ion transport, 117, 118, 127
Irradiation of phenolic compounds, 72

J

Jade plant, 71
Jasmonic acid, 36
Jerusalem artichoke,
ABA applications and, 9
auxins to increase hardiness and, 6
cerebroside as lipid component of plasma membrane and, 44
effects of ATPase activity and, 126
plasma membrane fluidity and, 47
sterols and, 42

K

Karyotypes, 41
Kharkov, see Winter wheat
Krebs cycle, 31
K^+ transport, 125—126

L

Laboratory of Plant Hardiness, 168
Lactate accumulation during flooding, 115—116
Lactic acid during flooding, 112
Lamellar-to-hexagomal II phase transition during freezing, 49
Lateral phase separation during freezing, 49
LCHI, 55, 57—58
LCHII, 55, 56—63
Legumes during ice encasement, 114
Lentils growth regulation, 71
Lettuce, 73, 83
Light, see Photoperiod entries; and Photosynthesis
Light dependency of *trans* 16:1 synthesis, 59, 60
Light exposure of ice encasement, 117
Light harvesting with LCHII, 55, 57—58
Light microscopy of protoplast osmotic behavior, 109
Light quality/sensitivity, 99
Light reactions with NADPH, 31
Light variables in cold hardening, 12, 185
Lignification during cold stress, 74—75
Lignin
effects of ABA and, 16
a phenolic biosynthesis product, 68, 70
premature germination inhibition and, 71

Linolenic acid production, 32, 34—35
Lipid acyl chain length, 124
Lipid bodies, 16
Lipid-exchange protein, 110
Lipid-protein interaction, 124
Lipids, see also Phospholipid entries
  biochemical changes in plasma membrane during freezing and, 42, 49
  metabolism of and, 36, 37
  physical properties of and, 44
  plasma membrane and, 54, 86, 109
  polyunsaturated fatty acids and, 31
  of thylakoid membrane, 54
  tonoplasts and, 42—44
  of winter wheat, 110
Lipoperoxide metabolism, 36
Liposomes, 49
Lipoxygenase, 36, 37
Liquid nitrogen, see LN
LN, 140—144, 147, 148
Low temperature exotherms, see LTE
LTE, 160—161
Lysine amino acids, 20
Lysolecithin, 131

M

Magnesium deficiency, 72
Malate, 30, 31, 37
Manitol, 144—146
Mass propagation of meristem/shoot tips, 140, 141, 143—144
Mechanical stress, 83
Mefluidide
  chilling injury protection and, 168—176
  concentrations and treatment intervals of and, 170, 171
  effects of ABA/GA and, 7, 173—175
  electrolyte leakage and, 172
  water content and, 171—172
Meristem/shoot tips
  mass propagation and, 140, 141, 143—144
  regrowth viability and, 149
Mesocotyl elongation, 11, 17
Messenger RNAs, see mRNA
Metabolic disequilibrium, 34—35
Metabolic equilibrium, 37
Metabolic modification
  assimilatory power and, 32—33
  ATP accumulation and, 33
  cell performance in low temperature and, 28
  during acclimation, 29—35
  first and second phases of and, 35—37
  plasma membrane permeability and, 36
  polyunsaturated fatty acid and, 31
Metabolic pathways, 32, 37
Metabolic shift, see Metabolic modification
Metabolic switch during growth cessation and promotion, 4
Metabolites, 31
Meteorological factors in forecasting, 160
Methionine, 178—179
Methyl transferases, 70
MGDG, 59, 60
MHL, 160—162
Microsomal fraction, 70
Microsomal phospholipids, 86
Mineral stress, 82
Minimum/maximum hardiness level, see MHL
Mitochondria
  ATP accumulation and, 34
  ATPase activity and, 131—132
  effects of ice encasement and, 116
  electron transport path and, 34
  oxidative phosphorylation and, 34
Mitochondrial membrane, 42
Molecular interactions between lipids and proteins, 44
Molecular motion/packing/rotation, 48—49
Monogalactosyldiacyl-glyceride, see MGDG
Monohydroxy phenolics, 73
Mono and oligomeric LCHII, 59, 60
Morphogenetic potentials, 141
Morphological alterations, 54
mRNA
  PAL synthesis and, 70
  polypeptides and, 16
  protein stimulation and, 9
Mulberry tree, 47
Multi-layered vesicles, 109
Mutant cell line, 142

N

NAD, 33
NADH, 35
NADH-cyt. c reductase, 134
NADP, 33
NADPH
  carbon assimilation pathways and, 31
  chloroplasts and, 31
  cold hardening and, 33
  dark reactions and, 31
  light reactions and, 31
  photosynthetic pathways and, 30
  pyridine nucleotide and, 31
  unsaturated lipids and phospholipids and, 32
Naringenin, 73
Nectarine, 162
Nitrogen deficiency, 72
Nitrogen nutrients, 11
NMR spectroscopy, 13
Northern ecotypes photoperiod variables, 92—94, 99
Northern grass species, 94—95, 104
Nuclear magnetic resonance, see NMR spectroscopy
Nucleation barrier of flower buds, 153—154
Nutrient stress, 70, 72

O

Oats, 185
Oligomeric LHCII, 57—63

Olive phenolic growth regulation, 71
Onion ATPase alterations, 125
o-Coumaric acid, 72
Optimal cooling rate of cryopreservation, 147
Orchard grass
plasma membrane fluidity and, 47
plasma membrane proteins and, 44—45
protein synthesis and, 16
sterols and, 42
Orientation of ATPase binding, 129—131
Orientation constraint, 49
Osmolality of cell sap, 16
Osmo regulation, 31—32, 37
Osmotic activity
cell membrane sensitivity and, 36
during freezing stress, 109
ethylene production and, 83
glucose and, 34
phases of metabolic modification and, 36
Osmotic strength, 9
Osmotic stress tolerance, 14—16
Overwintering crops, 115
Oxidative phosphorylation, 34
Oxygenation of polyunsaturated fatty acids, 36
Oxygen levels
effects of flooding and, 112
effects of ice encasement and, 115—118
Oxygen, 68

P

PAL
ammonia removal and, 68—70
disease resistance and, 72
ethylene stimulation and, 86
ethylene synthesis and, 73
photoalexin synthesis and, 82
Papaya, 182—183
Passionfruit, 72
p-coumaric acid
ethylene synthesis and, 73
PAL inhibition and, 70
phenolic compounds and, 71, 75
Peach
forecasting cold hardiness and, 159—162
GA's effect on cold hardening and, 159
LTE and, 162
phytoalexin induction and, 72
suberin and, 71
Peach flower bud properties, 153—156
Peanut/chickpea, 146—147
Pears, 73, 143—144
Peas, 7, 180
Pecan, 72
PEG 1000/PEG 6000, 145—146
Penetrating cryoprotective chemicals, 147
PEP carboxylase, 31
Perennial grass species, 92—95
Permeating/nonpermeating cryoprotectant, 145—146
Peroxidase
disease resistance and, 72
ethylene production and, 73, 82
role of in phenolic biosynthesis and, 70
Peroxide protection, 32
PG, 59—62
Phaseic acid, 17
Phases of cold hardening of woody plants, 102—104
Phases of cryopreservation, 148
Phase transition in inflection, 49
Phenolic biosynthesis, 68—71
Phenolic compounds, 16, 71—75
Phenylalanine, 68
Phenylaline ammonia-lyase, see PAL
Phophatidylethanolamine, see PTE losses
Phosopholipids, 31
Phospatidyl choline/glycerol, 185
Phosphate deficiency, 72
Phosphatidylcholine, 36
Phosphatidyldiacyl-glycerol, 54, see also PG
Phosphatidylethanolamine, 36
Phosphatidyl-glycerol, 36
Phosphatidyl transferase, 36
Phospholipase D, 36, 37
Phospholipid acyl chain catabolism, 32
Phospholipid catabolism, 37
Phospholipid component of plasma membrane, 86
Phospholipids, 32, 42, 47
Phospholipid-to-protein ratio, 110
Phosphorylation, 34
Photoalexin synthesis, 82, 86
Photochemical centers, 35
Photochemical reactions of thylakoid membrane, 54
Photomorphogenic effect, 92—93
Photooxidative degeneration, 185
Photoperiod of the Arctic ecotype, 92
Photoperiod changes, 4
Photoperiod variables
apical growth cessation and, 102
of the Arctic ecotype, 92
cold hardening of woody plants and, 98—99
leaf and stem elongation and, 92—94
Photorespiration, 35, see also Glycolic acid pathway
Photosynthesis, see also Respiration
biogenesis and, 57
effects of chilling injury under light and, 185
ion leakage and, 134
metabolism modification during cold hardening and, 29, 33—34
of reduced compounds, 30
thylakoid membrane protein complexes and, 110
Photosynthetic electron transport path, 30, 110, see also Electron transport path
Photosynthetic pathways, 30
Photosystem I, see PSI
Photosystem II, see PSII
Physical change, see Inflection
Phytoalexin induction, see specific plant species
Phytoalexins, 71—72
Phytochrome, 4, 6, 44
Pigment-protein complexes, 54, 55
Pine, 74
Pinus, 126

Pith injuries of woody plants, 101
Plant crown, 112
Plant regeneration, 141
Plasma membrane
  changes in composition of and, 42—49, 54, 86, 110
  effects of freezing injury and, 36, 54, 108, 109
Plasma membrane ATPase, see ATPase activity
Plasma membrane fluidity, 8—9, 47, 85
Plasma membrane integrity
  ATP accumulation and, 33
  effects of chilling injury and, 184—185
  proteins and, 44
Plasma membrane lipid bilayers, 44, 48
Plasma membrane lipids, 49, 125
Plasma membrane lysis, 109
Plasma membrane permeability
  chilling injury and, 172
  effects of ethylene production and, 86
  ion efflux and, 125
  metabolic modifications and, 36
Plasma membrane polypeptides, 44—45
Plasma membrane protein
  analysis by neutral detergent and, 45—46
  composition alterations and, 44—47
  inflection and, 49
  passive permeability and, 125
Plasma membrane purification, 127—131
Plasma membrane transport properties
  effects of ATPase activity and, 124, 127
  effects of ice encasement and, 117, 118
Polydisperse nonionic detergent, 46—47
Polyester components, 70
Polyethylene glycol, see PEG 1000/PEG 6000
Polymerization, 70
Polypeptide decomposition, 44—45
Polypeptides, see also Proteins
  as components of plasma membrane, 110
  as composition of soluble/insoluble proteins, 45—46
  effects of ABA and, 16—21
  hydrophobicity and, 46—47
  mRNA transcripts and, 16
  thylakoid membrane changes and, 54, 55
Polyphenol oxidase, 71—73
Polysomes, 9
Polyunsaturated fatty acids
  effects of chilling injury under light and, 185
  lipids/phospholipids and, 31
  metabolism modifications and, 31
  NADPH and ATP levels and, 32
Polyvinylpyrolidone, see PVP
Poplars, 6—7
Potassium deficiency, 72
Potato
  ABA applications and, 5, 9
  ATPase alterations during freeze injury and acclimation and, 124, 125, 127, 129, 130
  cryopreservation and, 146—147
  phytoalexin induction and, 72
  suberin and, 71
Preincubation in hypertonic solution, 147
Premature germination, 71
Primary induction, 93—94
Proliferation of spikelets, 94
Proline
  ABA applications and, 9
  amino acid accumulation and, 31—32
  cellular osmo regulation and, 31—32, 37
  as a cryoprotectant, 144—146
  growth during cold acclimation and, 30
Propyl gallate, 180
Prostrate growth habit, 17, 92
Protein, see also Polypeptides
  modification and, 29, 68
  photosynthesis and, 110
  in plasma membrane, 54
  of thylakoid membrane, 54
Protein-lipid interaction, 49, 135
Protein synthesis
  as affected by amino acid synthesis, 31—32
  cycloheximide and, 16
  effects of ABA in cold hardening and, 8—9, 16—21
  effects on ethylene production and, 179—180
  incubation conditions and, 17
  in winter rye, 16
Protein thiol groups, 32
Protocatechuic acid, 72
Proton uptake, 185
Protoplasts
  cryoprotection and, 146
  effects of ABA and, 16
  freezing stress and, 109
  plasma membrane and, 47
Prune, 85
PSI, 55, 57—58
PSII, 55, 57—58
PTE losses, 49
PVP 40,000, 145—146
Pyridine nucleotide
  cold hardening and, 31, 32
  glucose catabolism and, 33—34
  glucose metabolism and, 33—34
  NADPH and, 31
  redox state and, 33
Pyruvate kinase, 33—35

R

Radish, 8
Rapid freezing, 146—147
Raspberry, 143—144
Rb uptake, 127
Reaction phase of metabolic modification, 34—37
Reconstruction phase of metabolic modification, 34—36, see also Restitution phase of metabolic modification
Recrystalization, 148
Red light, 6, 103
Redox state of pyridine nucleotide levels, 33
Reduced compounds, 30

Reductive synthesis, 33
Rehydration, 110
Resistance to cellulase, 74—75
Respiration, 134, 184, see also Photosynthesis
Restitution phase of metabolic modification, 34—36, see also Reconstruction phase of metabolic modification
Rhizome hardening, 4—5
RH and RNH, 56—63
Rice, 168—176, 185
Ricinus communis, 8
Right-side-out ATPase orientation, 131
RNA, 9, 180
Rosetted growth habit, 92
Rotational relaxation time, 48—49
RuBP carboxylase, 31
Rubus chamaemorus L., see Cloudberry
Rye, 54, 55, see also Winter rye

S

S-adenosyl methionine, 84, see also SAM
Salinity, 30
Salix, see Woody plants
Salt stress, 14—16, 82
SAM
  effects on ethylene production and, 178—179, 185
  phenolic acid methylation and, 70
Saturation/unsaturation of plasma membrane bilayer lipids, 124
SDS gel electrophoresis, 56—57
SDS-polyacrylamide gel electrophoresis, 44—45
Second phase of metabolic modification, 35—37
Second stage of cold hardening, 85
Semipermeability of plasma membrane, 124
Sephadex gel column chromatography, 20
Serine, 30—32
Shikimic acid pathway derivative, 68
Shoot tips, 141, 143—144
Silver staining, 20
Slow cooling, 147
Slow freezing, 147
Smothering during ice encasement, 114, 115
Sodium benzoate, 180
Soil and air temperature variations during cold hardening, 4—5
Soil drainage, 110—119
Solanum, 126—135
Solute composition, 9
Sorbitol, 144—146
Southern ecotypes, 92—94, 99
Soybean
  ABA applications and, 8, 9
  effects of mefluidide on growth and, 170
  phytoalexin induction and, 72
Species tree line, 92
Spinach, 8, 9
Spin resonance, 59
Spring wheat, 33
SQDG, 59
Stability and organization changes of thylakoid membrane, 58
Starch grain size, 9
Starch granules, 16
Stem height/diameter, 83
Sterols, 42, 49, 110
Stoma of chloroplasts, 70
Stratification, 71
Strawberry, 146—147
Stress responses
  ethylene production and, 82—86, 178
  phenolic compounds and, 72—73
  plasma membrane proteins and, 44
Stress tolerance
  cross-adaptation and, 7—8, 14—16
  osmolality of cell sap and, 16
  water content and, 15, 16
Structural changes, 55—56, 110
Subculture, 141
Suberin, 68, 70—71
Suberization
  acclimation and, 70—71
  cold stress and, 74—75
  nutrient stress and, 72
Sucrose
  cold hardening and, 13
  as a cryoprotectant, 13, 144—146
  glycolic acid pathway and, 35
  phenolic accumulation and, 72
Sugar, see also Glucose
  the Calvin cycle and, 35
  metabolism modification during acclimation and, 29
  proteins in plasma membrane and, 46—47
Sugar cane, 9, 74
Sugar moieties, 46—47
Sulfoquinovoslyldiacyl-glyceride, see SQDG
Sulfur deficiency, 72
Sunflower, 8
Supercooling, 147, 148, 153
Supramolecular organization, 54, 60
Supramolecular pigment-protein complexes, 55
Sweet peppers, and ACC levels, 182
Sweet potato, 82, 184

T

TAL, 70
Tannin, 68, 71
TCA cycle, 35
Temperature fluctuation, 180—182
Temperature exotherm, 153
Terminal oxidase system, 35
Terpal, 7
Tetrameric enzyme, 70
Thermal properties of plasma membrane, 48
Thermotropic phase transition, 44
Thigmomorphogenesis, 70, 83
Thin layer chromatography, see TLC
Third stage of cold hardening, 85
Thylakoid membrane
  changes in structure and function and, 54—56

functions of and, 54—59
intrinsic and extrinsic components and, 55
lipids and, 54
membrane assembly and, 54
PG and, 59—62
supramolecular pigment-protein complexes and, 55
Thylakoid membrane pigment-protein complex, 54, 110
Thylakoids, 31, 185
Time variables, 11
Timothy, 94
Tissue regeneration, 143
Tissue structure of flower buds, 153—154
TLC, 73
Tobacco
ABA applications and, 8
cross adaptation to stress tolerance and, 7—8
cryopreservation and, 141
effects of light on ethylene production and, 185
phytoalexin induction and, 72
Tomato
ACC levels and, 184
cytokinins and, 7
ethylene production and, 83, 180
phenolic growth regulation and, 71
phytoalexin induction and, 72
suberin and, 71
Tonoplast contamination, 131—132
Tonoplasts, 42—44
Toxic chemicals, 82
Trans 16:1, 59—60
Trans-cinnamic acid, 68
Transhydrogenation reactions, 32
Translocatable hardiness inhibitors and promoters, 4—5
Transport properties of plasma membrane lipids, 125
Trans-Δ_3-hexadecanoic acid, see *Trans* 16:1
Triazoles, 7
Triphenyltetrazolium chloride, 149
Triton-soluble and insoluble proteins, 45—46
Triton X-114, 46—47
Typical browning, 73
Tyrosine, 68
Tyrosine ammonia-lyase, see TAL

U

Ultrastructure, 8—9
Ultra violet irradiation, 70
Ultra violet light absorption, 68
Uncouplers of oxidative phosphorylation, 180
University of Tromso, 92
Unsaturated fatty acids, 42, 47
Unsaturated lipids, 32
Unsaturated phospholipids, 42, 47

V

Vacuole changes, 16
Vapor pressure gradient, 147, 156
Variable photoperiods, 4—5
Vascular bundles, 71
Vascular development of flower buds, 154, 156
Vascular transition zone, 116
Vegetative buds, 101—104
Vernalization, 16
Virus infection, 70
Viviparous plants, 94, 104

W

Water content
effects on cold hardening northern grass species and, 94—95
stress tolerance and, 15, 16
Water flux, 8—9
Water loss, 13, 72
Water potential
effects of mefluidide and, 171
subfreezing pretreatment and, 36
supercooling of flower buds and, 155, 156
Water stress
ATP accumulation and, 33
biphasic physiologic response and, 34—35
effects of ABA levels and, 16, 171—173
ethylene production and, 82
osmo regulation and, 31—32
photosynthetic intermediates and, 30
Wheat, see also Winter wheat
effects of ethylene production and, 84
effects of ice encasement and, 112
phytoalexin induction and, 72
protein synthesis and, 16
stress resistance and, 83—84
White light, 6
White spruce, 71, 74
Winter barley, 108—110, 116, 117
Winter cereals
effects of ethylene production and, 84
effects of ice encasement and, 108—110, 113—119
freezing stress and, 108—110
injury resistance and, 108—119
Winter rape
ATP accumulation and, 33
cryopreservation and, 146—147
three stages of cold hardening and, 28—37
Winter rye, see also Rye
ABA to prevent chilling injury and, 5—6
analysis of plasma membrane proteins and, 44—45, 47
cell wall thickening and, 71
flooding stress and, 112
freezing stress and, 108—110
lignification and, 74
protein synthesis and, 16
sterols and, 42
Winter wheat
ABA levels and, 5—6
ATP accumulation and, 33
ATPase plasma membrane after acclimation and

freeze injury and, 126—135
cross adaptation to stress tolerance and, 7—8
cytokinins and, 7
effects of lipids on freezing stress and, 110
flooding stress and, 112
freezing stress and, 108—110, 116, 117
GA levels and, 6
lipids and, 110
prostrate growth habit and, 17
variable photoperiods and, 4
Woody perennials, 36
Woody plants
climatic variables affecting cold hardening and, 98—104
effects of ethylene production and, 84
growth behavior of and, 98—100
species improvement through cryopreservation and, 142—143
Wounding stress, 70, 82

X

Xanthone, 68, 69
Xylem vessels, 154, 156